T0234540

Lecture Notes in Computer Science   8920

Commenced Publication in 1973
Founding and Former Series Editors:
Gerhard Goos, Juris Hartmanis, and Jan van Leeuwen

More information about this series at http://www.springer.com/series/8637

Abdelkader Hameurlain · Josef Küng
Roland Wagner · Barbara Catania
Giovanna Guerrini · Themis Palpanas
Jaroslav Pokorný · Athena Vakali (Eds.)

Transactions on Large-Scale Data- and Knowledge-Centered Systems XV

Selected Papers from ADBIS 2013
Satellite Events

 Springer

Editors-in-Chief

Abdelkader Hameurlain
IRIT
Paul Sabatier University
Toulouse
France

Roland Wagner
FAW
University of Linz
Linz
Austria

Josef Küng
FAW
University of Linz
Linz
Austria

Guest Editors

Barbara Catania
University of Genoa
Genoa
Italy

Jaroslav Pokorný
Charles University
Prague
Czech Republic

Giovanna Guerrini
University of Genoa
Genoa
Italy

Athena Vakali
Aristotle University of Thessaloniki
Thessaloniki
Greece

Themis Palpanas
Paris Descartes University
Paris
France

ISSN 0302-9743   ISSN 1611-3349   (electronic)
Lecture Notes in Computer Science
ISBN 978-3-662-45760-3   ISBN 978-3-662-45761-0   (eBook)
DOI 10.1007/978-3-662-45761-0

Library of Congress Control Number: 2014957888

Springer Heidelberg New York Dordrecht London

Printed on acid-free paper

Springer-Verlag GmbH Berlin Heidelberg is part of Springer Science+Business Media
(www.springer.com)

Preface
Special Issue: Selected Papers from ADBIS 2013
Satellite Events

The 17th East-European Conference on Advances in Databases and Information Systems (ADBIS 2013) took place in Genoa, Italy, from September 1 to 4, 2013. The ADBIS series of conferences aims at providing a forum for the dissemination of research accomplishments and at promoting interaction and collaboration between the database and information systems research communities from Central and East European countries and the rest of the world. The ADBIS conferences provide an international platform for the presentation of research on database theory, development of advanced DBMS technologies, and their advanced applications. As such, ADBIS has meanwhile created a tradition: its 2013 edition continued the ADBIS series held in St. Petersburg (1997), Poznań (1998), Maribor (1999), Prague (2000), Vilnius (2001), Bratislava (2002), Dresden (2003), Budapest (2004), Tallinn (2005), Thessaloniki (2006), Varna (2007), Pori (2008), Riga (2009), Novi Sad (2010), Vienna (2011), Poznań (2012). The program of the 2013 edition included keynotes, research papers, and five satellite events, consisting of a Big Data special session, three thematic workshops, and a Doctoral Consortium.

In the present special issue, the extended and revised version of four papers out of the twenty-five papers presented at ADBIS 2013 Satellite Events are included. The papers cover various topics in large-scale data- and knowledge-centered systems, including GPU-accelerated database systems and GPU-based compression for large time series databases, design of parallel data warehouses, and schema matching.

The first two papers were presented at GID 2013 – Second International Workshop on GPUs in Databases.

The first paper, by Sebastian Breß, Max Heimel, Norbert Siegmund, Ladjel Bellatreche, and Gunter Saake, is entitled "GPU-Accelerated Database Systems: Survey and Open Challenges" and explores the design space of GPU-accelerated database management systems. Based on the proposed survey, key properties and typical challenges of GPU-aware database architectures are presented and open research problems are formulated. Existing GPU-accelerated database management systems are also surveyed and their architectural properties classified, with a special emphasis on optimization issues. A reference architecture is finally proposed, indicating how GPU acceleration can be integrated in existing DBMSs.

The second paper is entitled "Compression Planner for Time Series Database with GPU Support" and is co-authored by Piotr Przymus and Krzysztof Kaczmarski. The paper exploits GPU in designing a compression planner for time series databases. Motivated by the fact that the growing volumes of time series data call for the definition of efficient and innovative processing approaches, the paper presents a novel compression method which is ultra fast and achieves the best possible compression ratio by composing several lightweight algorithms dynamically tuned for incoming data. The reported experimental results show that the proposed approach is a valid solution for data intensive monitoring and analytic systems.

The third paper, presented at SoBI 2013 – First International Workshop on Social Business Intelligence: Integrating Social Content in Decision Making, is entitled "A Global Paradigm for Designing Parallel Relational Data Warehouses in Distributed Environments" and is co-authored by Soumia Benkrid, Ladjel Bellatreche, and Alfredo Cuzzocrea. The paper proposes a novel methodology for designing a Parallel Relational Data Warehouse, in which, differently from other existing proposals, all the main design phases (i.e., fragmentation, allocation, and replication) are performed simultaneously, in a global fashion. The reported experimental results assess the performance of the proposed methodology against a well-known data warehouse benchmark.

The fourth paper, by Alsayed Algergawy, Seham Moawed, Amany Sarhan, Ali Eldosouky, and Gunter Saake, is entitled "Improving Clustering-Based Schema Matching Using Latent Semantic Indexing" and was presented at OAIS 2013 – Second International Workshop on Ontologies Meet Advanced Information Systems. The paper focuses on the identification of semantically corresponding elements across heterogeneous and large datasets and proposes a clustering-based matching algorithm which relies on a Latent Semantic Indexing-based approach in order to guarantee a high quality yet efficient process. Such an approach allows the retrieval of the conceptual meaning between clusters and the identification of the hidden semantic relationships among clusters elements. Experimental results are also reported, showing that the proposed approach is quite promising.

Several people have contributed to making this special issue possible. We are grateful to Abdelkader Hameurlain, Josef Küng, and Roland Wagner, Editors-in-Chief of the TLDKS journal, for accepting our proposal for this special issue. Special thanks are due to Gabriela Wagner, for her valuable assistance during the preparation of this special issue. We would also like to express our appreciation to the authors who submitted their manuscripts to be considered for this special issue. Last but not least, we would like to gratefully acknowledge the contribution of all the reviewers, who worked within a very tight schedule and whose detailed and constructive feedback to the authors contributed to improving the quality of the submitted manuscripts.

November 2014

Barbara Catania
Giovanna Guerrini
Themis Palpanas
Jaroslav Pokorný
Athena Vakali

Editorial Board

Contents

GPU-Accelerated Database Systems: Survey and Open Challenges

Sebastian Breß[1]([✉]), Max Heimel[2], Norbert Siegmund[3],
Ladjel Bellatreche[4], and Gunter Saake[1]

[1] University of Magdeburg, Magdeburg, Germany
{sebastian.bress,gunter.saake}@ovgu.de
[2] Technische Universität Berlin, Berlin, Germany
max.heimel@tu-berlin.de
[3] University of Passau, Passau, Germany
siegmunn@fim.uni-passau.de
[4] LIAS/ISAE-ENSMA, Futuroscope, Poitiers, France
bellatreche@ensma.fr

Abstract. The vast amount of processing power and memory bandwidth provided by modern graphics cards make them an interesting platform for data-intensive applications. Unsurprisingly, the database research community identified GPUs as effective co-processors for data processing several years ago. In the past years, there were many approaches to make use of GPUs at different levels of a database system. In this paper, we explore the design space of GPU-accelerated database management systems. Based on this survey, we present key properties, important trade-offs and typical challenges of GPU-aware database architectures, and identify major open challenges. Additionally, we survey existing GPU-accelerated DBMSs and classify their architectural properties. Then, we summarize typical optimizations implemented in GPU-accelerated DBMSs. Finally, we propose a reference architecture, indicating how GPU acceleration can be integrated in existing DBMSs.

Keywords: GPU-accelerated database · Survey · Co-processing · Modern database architecture

1 Introduction

Over the last few years, the traditional performance drivers of modern processors – frequency and parallelism – started to hit physical limits. One reason for this is that modern processors are constrained to a certain amount of power they may consume (i.e., the power wall [12]) and further increasing frequency and parallelism would make them overly power hungry. Therefore, hardware

---

This paper is a substantially extended version of an earlier work [17].

© Springer-Verlag Berlin Heidelberg 2014
A. Hameurlain et al. (Eds.): TLDKS XV, LNCS 8920, pp. 1–35, 2014.
DOI: 10.1007/978-3-662-45761-0_1

vendors are forced to create processors that are optimized for a certain application field. These developments result in a highly heterogeneous hardware landscape, which is expected to become even more diverse in the future [12]. In order to keep up with the performance requirements of the modern information society, tommorow's database systems will need to exploit and embrace this increased heterogeneity.

In this article, we take a closer look at how today's database engines manage heterogeneous environments, demonstrated by systems that support *Graphics Processing Units* (GPUs). The GPU is the pioneer of modern co-processors, and – in the last decade – it matured from a highly specialized processing device to a fully programmable, powerful co-processor. This development inspired the database research community to investigate methods for accelerating database systems via GPU co-processing. Several research papers and performance studies demonstrate the potential of this approach [7, 21, 29, 32, 48, 49] – and the technology has also found its way into commercial products (e.g., Jedox [1] or ParStream [2]).

Using graphics cards to accelerate data processing is tricky and has several pitfalls: First, for effective GPU co-processing, the transfer bottleneck between CPU and GPU has to either be reduced or concealed via clever data placement or caching strategies. Second, when integrating GPU co-processing into a real-world *Database Management System* (DBMS), the challenge arises that DBMS internals – such as data structures, query processing and optimization – are traditionally optimized for CPUs. While there is ongoing research on building GPU-aware database systems [22], no unified GPU-aware DBMS architecture has emerged so far.

In this paper, we want to make the community aware of the lack of a unified GPU-aware architecture and derive – based on a literature survey – a reduced design space for such an architecture. In particular, we make the following contributions:

1. We traverse the design space for a GPU-aware database architecture based on results of prior work.
2. We derive research questions that should be investigated by the community to develop GPU-aware database architectures.

Furthermore, as a substantial extension to a previous version of this paper [17], we conducted an in-depth literature survey of eight GPU-accelerated database management systems to validate and refine our theoretical discussions. This complements our findings in proposing a reference architecture. In detail, we make the following additional contributions:

1. We discuss eight *GPU-accelerated DBMSs* (GDBMSs) to review the state-of-the-art, collect prominent findings, and complement our discussion on a GPU-aware DBMS architecture.
2. We create a classification of required architectural properties of GDBMSs.
3. We summarize optimizations implemented by the surveyed systems and derive a general set of optimizations that a GDBMS should implement.

4. We propose a reference architecture for GDBMSs. This architecture provides insights on how to integrate GPU acceleration in main-memory DBMSs.
5. We identify new open challenges compared to our earlier work [17].

We find that GDBMSs should be in-memory column stores, should use the block-at-a-time processing model and exploit all available processing devices for query processing by using a GPU-aware query optimizer. Thus, main memory DBMSs are similar to GPU-accelerated DBMSs, and most in-memory, column-oriented DBMSs can be extended to efficiently support co-processing on GPUs.

The paper is structured as follows: In Sect. 2, we provide necessary background information about GPUs and discuss related work. We explore the design space for GPU-accelerated DBMSs w.r.t. functional and non-functional properties in Sect. 3. In Sect. 4, we survey a representative set of GPU-accelerated DBMSs, classify their architectural properties, summarize possible optimizations to speed up query processing, and propose a reference architecture for GDBMSs. Finally, we identify open challenges for GDBMSs in Sect. 5 and summarize our findings in Sect. 6.

2 Preliminary Considerations

In this section, we provide a brief overview over the architecture of graphics cards, the applied programming model, and related work.

2.1 Graphics Card Architecture

Figure 1 shows the architecture of a modern computer system with a graphics card. The figure shows the architecture of a graphics card from the *Tesla* architecture of NVIDIA. While specific details might be different for other vendors, the general concepts are found in all modern graphic cards. The graphics card – henceforth also called the *device* – is connected to the *host system* via the *PCIExpress bus*. All data transfer between host and device has to pass through this comparably low-bandwidth bus.

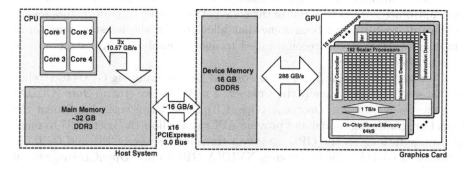

Fig. 1. Overview: Exemplary architecture of a system with a graphics card.

The graphics card itself contains one or more GPUs and a few gigabytes of *device memory*.[1] Typically, host and device do not share the same address space, meaning that neither the GPU can directly access the main memory nor the CPU can directly access the device memory.

The GPU itself consists of a few *multiprocessors*, which can be seen as very wide SIMD processing elements. Each multiprocessor packages several *scalar processors* with a few kilobytes of high-bandwidth, on-chip *shared memory*, cache, and an interface to the device memory.

2.2 Programming a GPU

Programs that run on a graphics card are written in the so-called *kernel programming model*. Programs in this model consist of *host code* and *kernels*. The host code manages the graphics card, initializing data transfer and scheduling program execution on the device. A kernel is a simplistic program that forms the basic unit of parallelism in the kernel programming model. Kernels are scheduled concurrently on several scalar processors in a SIMD fashion: Each kernel invocation - henceforth called *thread* - executes the same code on its own share of the input. All threads that run on the same multiprocessor are logically grouped into a *workgroup*.

One of the most important performance factors in GPU programming is to avoid data transfers between host and device: All data has to pass across the PCIexpress bus, which is the bottleneck of the architecture. Data transfer to the device might therefore consume all time savings from running a problem on the GPU. This becomes especially evident for I/O-bound algorithms: Since accessing the main memory is roughly two to three times faster than sending data across the PCIexpress bus, the CPU will usually have finished execution before the data has even arrived on the device.

Graphics cards achieve high performance through massive parallelism. This means, that a problem should be easy to parallelize to gain most from running on the GPU. Another performance pitfall in GPU programming is caused by divergent code paths. Since each multiprocessor only has a single instruction decoder, all scalar processors execute the same instruction at a time. If some threads in a workgroup diverge, for example due to data-dependent conditionals, the multiprocessor has to serialize the code paths, leading to performance losses. While this problem has been somewhat alleviated in the latest generation of graphics cards, it is still recommended to avoid complex control structures in kernels where possible.

Currently, two major frameworks are used for programming GPUs to accelerate database systems, namely the *Compute Unified Device Architecture* (CUDA) and the *Open Compute Language* (OpenCL). Both frameworks implement the kernel programming model and provide API's that allow the host CPU to manage computations on the GPU and data transfers between CPU and GPU. In contrast to CUDA, which supports NVIDIA GPUs only, OpenCL can run on

---

[1] Typically around 2–4 GB on mainstream cards and up to 16 GB on high-end devices.

a wide variety of devices from multiple vendors [24]. However, CUDA offers advanced features such as allocation of device memory inside a running kernel or *Uniform Virtual Addressing* (UVA), a technique where CPUs and GPUs share the same virtual address space and the CUDA driver transfers data between CPU and GPU transparently to the application [45].[2]

2.3 Related Work

To the best of our knowledge, there is no survey summarizing the state-of-the-art of GPU-accelerated DBMSs. The only survey we are aware of is from Owens and others, which discusses the state-of-the-art in GPGPU computing [46]. They cover a wide area of research, mainly GPGPU techniques (e.g., stream operations, data structures, and data queries) and GPGPU applications (e.g., databases and data mining, physically-based simulation, and signal and image processing). In contrast to Owens, we focus on recent trends in GPU-accelerated data management to derive a GPU-aware database architecture and open research questions.

3   Exploring the Design Space of a GPU-Aware DBMS Architecture

In this section, we explore the design space of a GPU-accelerated database management system from two points of views: Non-functional properties (e.g., performance and portability) and functional properties (e.g., transaction management and processing model). Note that while we focus on relational systems, most of our discussions apply to other data models as well.

3.1   Non-functional Properties

In the following, we discuss non-functional properties for which DBMSs are typically optimized for, namely performance and portability, and the introduced problems when supporting GPUs. Tsirogiannis and others found that in most cases, the configuration performing best is also the most energy efficient configuration due to the large up-front power consumption in modern servers [59]. Therefore, we will not discuss energy efficiency separately, as energy efficiency is already covered by the performance property.

Performance. Since the GPU is a specialized processor, it is faster on certain tasks (e.g., numerical computations) than CPUs, whereas CPUs outperform GPUs for tasks that are hard to parallelize or that involve complex control flow instructions. He and others observed that joins are 2–7 times faster on the GPU, whereas selections are 2–4 times slower, due to the required data transfers [30].

---

[2] We are aware that this features are included in OpenCL 2.0 but no OpenCL framework supports this features yet.

The same conclusion was made by Gregg and others, who showed that a GPU algorithm is not necessarily faster than its CPU counterpart, due to the expensive data transfers [27]. One major point for achieving good performance in a GDBMS is therefore to avoid data transfers where possible.

Another problem is how to select the optimal processing device for a given operation. For instance: While the GPU is well suited for easily parallelizable operations (e.g., predicate evaluation, arithmetic operations), the CPU is the vastly better fit when it comes to operations that require complex control structures or significant inter-thread communications (e.g., hash table creation or complex user-defined functions). Selecting the optimal device for a given operation is a non-trivial operation, and – due to the large parameter space (e.g., Breß and others [14] or He and others [29]) – applying simple heuristics is typically insufficient. Breß and others argue that there are four major factors that need to be considered for such a decision (1) the operation to execute, (2) the features of the input data (e.g., data size, data type, operation selectivity, data skew), (3) the computational power and capabilities of the processing devices (e.g., number of cores, memory bandwidth, clock rate), and (4) the load on the processing device (e.g., even if an operation is typically faster on the GPU, one should use the CPU when the GPU is overloaded) [14]. Therefore, we argue that a complex decision model, that incorporates these four factors, is needed to decide on an optimal operator placement.

Portability. Modern DBMSs are tailored towards CPUs and apply traditional compiler techniques to achieve portability across the different CPU architectures (e.g., x86, ARM, Power). By using GPUs – or generally, heterogeneous coprocessors – this picture changes, as CPU code cannot be automatically ported to run efficiently on a GPU. Also, certain GPU toolkits – such as CUDA – bind the DBMS vendor to a certain GPU manufacturer.

Furthermore, processing devices themselves are becoming more and more heterogeneous [55]. In order to achieve optimal performance, each device typically needs its own optimized version of the database operators [19]. However, this means that supporting all combinations of potential devices yields an exponential increase in required code paths, leading to a significant increase in development and maintenance costs.

There are two possibilities to achieve portability also for GPUs: First, we can implement all operators for all vendor-specific toolkits. While this has the advantage that special features of a vendor's product can be used to achieve high performance, it leads to high implementation effort and development costs. Examples for such systems are GPUQP [29] or CoGaDB [13], a column-oriented and GPU-accelerated DBMS. Second, we can implement the operators in a generic framework, such as OpenCL, and let the hardware vendor provide the optimal mapping to the given GPU. While this approach saves implementation effort and simplifies maintenance, it also suffers from performance degradation compared to hand- tuned implementations frameworks. To the best of our knowledge, the only system belonging to the second class is Ocelot [34], which extends MonetDB with OpenCL-based operators.

Summary. From the discussion, it is clearly visible that GPU acceleration complicates the process of optimizing GDBMSs for non-functional properties such as performance and portability. Thus, we need to take special care to achieve comparable applicability with respect to traditional DBMSs.

3.2   Functional Properties

We now discuss the design space for a relational GDBMS with respect to functional properties. We consider the following design decisions: (1) main-memory vs. disk-based system, (2) row-oriented vs. column-oriented storage, (3) processing models (tuple-at-a-time model vs. operator-at-a-time), (4) GPU-only vs. hybrid device database, (5) GPU buffer management (column-wise or page-wise buffer), (6) query optimization for hybrid systems, and (7) consistency and transaction processing (lock-based vs. lock free protocols).

Main-Memory vs. Hard-Disk-Based System. He and others demonstrated that GPU-acceleration cannot achieve significant speedups if the data has to be fetched from disk, because of the IO bottleneck, which dominates execution costs [29]. Since the GPU improves performance only once the data has arrived in main memory, time savings will be small compared to the total query runtime. Hence, a GPU-aware database architecture should make heavy use of in-memory technology.

Row-Stores vs. Column Stores. Ghodsnia compares row and column stores with respect to their suitability for GPU-accelerated query processing [25]. Ghodsnia concluded that a column store is more suitable than a row store, because a column store (1) allows for coalesced memory access on the GPU, (2) achieves higher compression rates (an important property considering the current memory limitations of GPUs), and (3) reduces the volume of data that needs to be transferred. For example, in case of a column store, only those columns needed for data processing have to be transferred between processing devices. In contrast, in a row-store, either the full relation has to be transferred or a projection has to reduce the relation to the data needed to process a query. Both approaches are more expensive than storing the data column wise. Bakkum and others came to the same conclusion [6]. Furthermore, given that we already concluded that a GPU-aware DBMS should be an in-memory database system, and that current research provides an overwhelming evidence in favor of columnar storage for in-memory systems [10]. We therefore conclude that a GPU-aware DBMS should use columnar storage.

Processing Model. There are basically two alternative processing models that are used in modern DBMS: the tuple-at-a-time model [26] and operator-at-a-time bulk processing [42]. Tuple-at-a-time processing has the advantage that intermediate results are very small, but has the disadvantage that it introduces a higher per tuple processing overhead as well as a high cache miss rate. In contrast, operator-at-a-time processing leads to cache friendly memory access patterns, making effective usage of the memory hierarchy. However, the major drawback is the increased memory requirement, since intermediate results are materialized [42].

Tuple-at-a-time approaches usually apply the so-called *iterator model*, which applies virtual function calls to pass tuples through the required operators [26]. Since graphics cards lack support for virtual function calls – and are notoriously bad at runing the complex control logic that would be necessary to emulate them – this model is unsuited for a GDBMS. Furthermore, we identified in prior work that tuple-wise processing is not possible on the GPU, due to lacking support for inter-kernel communication [15]. We therefore argue that a GDBMS should utilize an operator-at-a-time model.

In order to avoid the IO overhead of this model, multiple authors have suggested a hybrid strategy that uses dynamic code compilation to merge multiple logical operators, or even express the whole query in a single, runtime-generated operator [20,44,60]. Using this strategy, it is not necessary to materialize intermediate results in the GPU's device memory: Tuples are passed between operators in registers, or via shared memory. This approach is therefore an additional potential execution model for a GDBMS.

Database in GPU RAM vs. Hybrid Device Database. Ghodsnia proposed to keep the complete database resident in GPU RAM [25]. This approach has the advantage of vastly reducing data transfers between host and device. Also, since the GPU RAM has a bandwidth that is roughly 16 times higher than the PCIe Bus (3.0), this approach is very likely to significantly increase performance. It also simplifies transaction management, since data does not need to be kept consistent between CPU and GPU.

However, the approach has some obvious shortcomings: First, the GPU RAM (up to ≈16 GB) is rather limited compared to CPU RAM (up to ≈2 TB), meaning that either only small data sets can be processed, or that data must be partitioned across multiple GPUs. Second, a pure GPU database cannot exploit full inter-device parallelism, because the CPU does not any perform data processing. Since CPU and GPU both have their corresponding sweet-spots for different applications (cf. Sect. 3.1), this is a major shortcoming that significantly degrades performance in several scenarios.

Since these problems outweigh the benefits, we conclude that a GDBMS should make use of all available storage and not constrain itself to GPU RAM. While this complicates data processing, and requires a data-placement strategy[3], we still expect the hybrid to be faster than a pure CPU- or GPU-resident system. The performance benefit of using both CPU and GPU for processing was already

---

[3] Some potential strategies include keeping the hot set of the data resident on the graphics card, or using the limited graphics card memory as a low-resolution data storage to quickly filter out non-matching data items [47].

observed for hybrid query processing approaches (e.g., He and others [29] and Breß and others [18]).

Effective GPU Buffer Management. The buffer-management problem in a CPU/GPU system is similar to the one encountered in traditional disk-based or in-memory systems. That is, we want to process data in a faster, and smaller memory space (GPU RAM), whereas the data is stored in a larger and slower memory space (CPU RAM). The novelty in this problem is, that – in contrast to traditional systems – data can be processed in both memory spaces. In other words: We can transfer data, but we do not have to! This *optionality* opens up some interesting research questions, that have not been covered in traditional database research.

Data structures and data encoding are often highly optimized for the special properties of a processing device to maximize performance. Hence, different kinds of processing devices use an encoding optimized for the respective device. For example, a CPU encoding has to support effective caching to reduce the memory access cost [41], whereas a GPU encoding has to ensure coalesced memory access of threads to achieve maximal performance [45]. This usually requires transcoding data before or after the data transfer, which is an additional overhead that can break performance.

Another interesting design decision is the granularity that should be used for managing the GPU RAM: pages, whole columns, or whole tables? Since we already concluded that a GPU-accelerated database should be columnar, this basically boils down to comparing page-wise vs. column-based caching. Page-wise caching has the advantage that it is an established approach, and is used by almost every DBMS, which eases integration into existing systems. However, a possible disadvantage is that – depending on the page size –, the PCIe bus may be underutilized during data transfers. Since it is more efficient to transfer few large data sets than many little datasets (with the same total data volume) [45], it could be more beneficial to cache and manage whole columns.

Query Placement and Optimization. Given that a GPU-aware DBMS has to manage multiple processing devices, a major problem is to automatically decide which parts of the query should be executed on which device. This decision depends on multiple factors, including the operation, the size & shape of the input data, processing power and computational characteristics of CPU and GPU as well as the optimization criterion. For instance: Optimizing for response time requires to split a query in parts, so that CPU and GPU can process parts of the query in parallel. However, workloads that require a high throughput, need different heuristics. Furthermore, given that we can freely choose between multiple different processing devices with different energy characteristics, non-traditional optimization criteria such as energy-consumption, or cost-per-tuple become interesting in the scope of GPU-aware DBMSs.

He and others were the first to address hybrid CPU/GPU query optimization [29]. They used a Selinger-style optimizer to create initial query plans and then used heuristics and an analytical cost-model to split a workload between CPU and GPU. In our previous work, we proposed a framework that can perform cost-based operation-wise scheduling and cost-based optimization of hybrid CPU/GPU query plans, which is designed to be used with operator-at-a-time bulk processing [15]. Przymus and others developed a query planner that is capable of optimizing for two goals simultaneously (e.g., query response time and energy consumption) [51]. Heimel and others suggest using GPUs to accelerate query optimization instead of query processing. This approach could help to tackle the additional computational complexity of query optimization in a hybrid system [33]. It should be noted that there is some similarity to the problem of query optimization in the scope of distributed and federated DBMSs [39]. However, there are several characteristics that differentiate distributed from hybrid CPU/GPU query processing:

1. In a distributed system, nodes are autonomous. This is in contrast to hybrid CPU/GPU systems, because the CPU has to explicitly send commands to the co-processors.
2. In a distributed system, there is no global state. By contrast, in hybrid CPU/GPU systems the CPU knows which co-processor performs a certain operation on a specific dataset.
3. The nodes in a distributed system are loosely coupled, meaning that a node may loose network connectivity to the other nodes or might crash. In a hybrid CPU/GPU system, nodes are tightly bound. That is, no network outages are possible due to a high bandwidth bus connection, and a GPU does not go down due to a local software error.

We conclude that traditional approaches for a distributed system do not take into account specifics of hybrid CPU/GPU systems. Therefore, tailor-made co-processing approaches are likely to outperform approaches from distributed or federated query-processing.

Consistency and Transaction Processing. Keeping data consistent in a distributed database is a widely studied problem. But, research on transaction management on the GPU is almost non-existent. The only work we are aware of is by He and others [31] and indicates that a locking-based strategy significantly breaks the performance of GPUs [31]. They developed a lock-free protocol to ensure conflict serializability of parallel transactions on GPUs. However, to the best of our knowledge, there is no work that explicitly addresses transaction management in a GDBMS. It is therefore to be investigated how the performance characteristics of established protocols of distributed systems compare to tailor-made transaction protocols.

Essentially, there are three ways of maintaining consistency between CPU and GPU: (1) Each data item could be kept strictly in one place (e.g., using horizontal or vertical partitioning). In this case, we would not require any replication management and would have to solve a modified allocation problem. (2) We can use established replication mechanisms, such as read one write all or

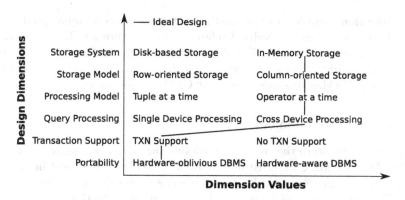

Fig. 2. Design space of GPU-aware DBMSs

primary copy. (3) The system can perform updates always on one processing device (e.g., the CPU) and periodically synchronize these changes to the other devices.

3.3 Summary

We summarize the results of our theoretical discussion in Fig. 2. A GPU-aware database system should reside in-memory and use columnar storage. As processing model, it should implement operator-at-a-time bulk processing model, potentially enhanced by dynamic code compilation. The system should make use of all available (co-)processors in the system (including the CPU!) by having a locality-aware query optimizer, which distributes the workload across all available processing resources. In case the GPU-aware DBMS needs transaction support, it should use an optimistic transaction protocol, such as the timestamp protocol. Finally, in order to reduce implementation overhead, the ideal GDBMS would be hardware-oblivious, meaning all hardware-specific adaption is handled transparently by the system itself.

While this theoretical discussion already gave us a good idea of how the reference architecture for a GDBMS should look like, we will now take a closer look at existing GDBMSs to refine our results.

4 A Survey of GPU-Accelerated DBMSs

In this section, we refine our theoretical discussion of the GDBMS design space from Sect. 3 by conducting a survey on existing GPU-accelerated database systems. First, we describe our research methodology. Second, we discuss the architectural properties of all systems that meet our survey selection criteria. Third, we classify the systems according to our design criteria (cf. Sect. 3). Based on our classification, we then discuss further optimization techniques used in the surveyed systems. Then, we derive a reference architecture for GPU-accelerated DBMSs based on our results. Finally, we will use this reference architecture for GDBMSs to identify a set of extensions that is required to extend existing main-memory DBMSs to support efficient GPU co-processing.

4.1   Research Methodology

In this section, we state the research questions that drive our survey. Then, we describe the selection criteria to find suitable DBMS architectures in the field of GPU-acceleration. Afterwards, we discuss the properties we focus on in our survey. This properties will be used as base for our classification.

Research Questions

RQ1: Are there recurring architectural properties among the surveyed systems?
RQ2: Are there application-specific classes of architectural properties?
RQ3: Can we infer a reference architecture for GPU-accelerated DBMSs based on existing GPU-accelerated DBMSs?
RQ4: How can we extend existing main-memory DBMSs to efficiently support data processing on GPUs?

Selection Criteria. Since this survey should cover relational GDBMS, we only consider systems that are capable of using the GPU for most relational operations. That is, we disregard stand-alone approaches for accelerating a certain relational operator (e.g., He and others [30,32]), special co-processing techniques (e.g., Pirk and others [49]), or other – non data-processing related – applications for GPUs in database systems [33]. Furthermore, we will not discuss systems using other data models than the relational model, such as graph databases (e.g., Medusa from Zhong and He [64,65]) or MapReduce (e.g., Mars from He and others [28]). Also, given that publications, such as research papers or whitepapers, often lack important architectural informations, we strongly preferred systems that made their source code publicly available. This allowed us to analyze the source code in order to correctly classify the system.

Comparison Properties. According to the design decisions discussed in Sect. 3, we present for each GDBMS the storage system, the storage and processing model, query placement and query optimization, and support for transaction processing. The information for this comparison is taken either directly from analyzing the source code – if available –, or from reading through published articles about the system. If a properties is not applicable for a system, we mark it as not applicable and focus on unique features instead.

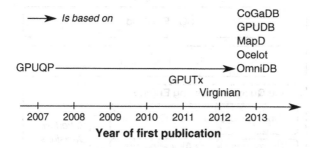

Fig. 3. Time line of surveyed systems.

4.2   GPU-Accelerated DBMS

Based on the discussed selection criteria, we identified the following eight academic[4] systems that are relevant for our survey:

System	Institute	Year	Open Source	Ref.
CoGaDB	University of Magdeburg	2013	yes	[13,18]
GPUDB	Ohio State University	2013	yes	[62]
GPUQP	Hong Kong University of Science and Technology	2007	yes	[29]
GPUTx	Nanyang Technological University	2011	no	[31]
MapD	Massachusetts Institute of Technology	2013	no	[43]
Ocelot	Technische Universität Berlin	2013	yes	[34]
OmniDB	Nanyang Technological University	2013	yes	[63]
Virginian	NEC Laboratories America	2012	yes	[6]

In Fig. 3, we illustrate the chronological order in which the first publications for each system were published. It is clearly visible that most systems were developed very recently and only few systems are based on older systems. Hence, we expect little influence on the concrete DBMS architecture between each other and hence, a strong external validity of our results.

CoGaDB

Breß and others developed a column-oriented GPU-accelerated DBMS (CoGaDB[5]) [13,18]. CoGaDB focuses on GPU-aware query optimization to achieve efficient co-processor utilization during query processing (Fig. 4).

---

[4] Note that we deliberately excluded commercial systems such as Jedox [1] or Parstream [2], because they are neither available as open source nor have publications available that provide full architectural details.

[5] Source code available at: http://wwwiti.cs.uni-magdeburg.de/iti_db/research/gpu/cogadb/.

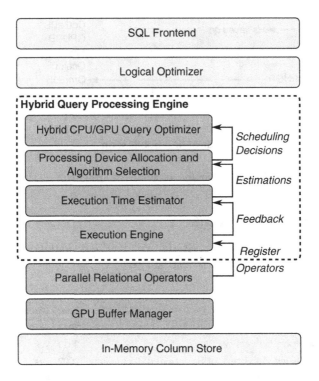

Fig. 4. The architecture of CoGaDB, taken from [16]

Storage System: CoGaDB persists data on disk, but loads the complete database into main memory on startup. If the database is larger than the main memory, CoGaDB relies on the operating system's virtual memory management to swap the least recently used memory pages on disk.

Storage Model: CoGaDB stores data in data structures optimized for in-memory databases. Hence, it stores the data column-wise and compresses VARCHAR columns using dictionary encoding [9]. Furthermore, the data has the same format when stored in the CPU's or the GPU's memory.

Processing Model: CoGaDB uses the operator-at-a-time bulk processing model to make efficient use of the memory hierarchy. This is the basis for efficient query processing using all processing resources.

Query Placement & Optimization: CoGaDB uses the *Hybrid Query Processing Engine* (HyPE) as physical optimizer [13]. HyPE optimizes physical query plans to increase inter-device parallelism by keeping track of the load condition on all (co-)processors (e.g., the CPU or the GPU).

Transactions: Not supported.

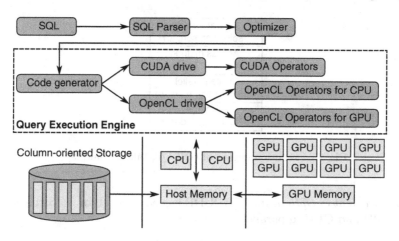

Fig. 5. GPUDB: Query engine architecture, taken from [62]

GPUDB

In order to study the performance behaviour of OLAP queries on GPUs, Yuan and others developed GPUDB[6] [62] (Fig. 5).

Storage System: GPUDB keeps the database in the CPU's main memory to avoid the hard-disk bottleneck. Yuan and others identified a crucial optimization for main-memory DBMS with respect to GPU accelerated execution: In case data is stored in pinned host memory, query execution times can significantly improve (i.e., Yuan and others observed speedups up to 6.5x for certain queries of the *Star Schema Benchmark* (SSB) [52]).

Storage Model: GPUDB stores the data column-wise because GPUDB is optimized for warehousing workloads. Additionally, GPUDB supports common compression techniques (run length encoding, bit encoding, and dictionary encoding) to decrease the impact of the PCIe bottleneck and to accelerate data processing.

Processing Model: GPUDB uses a block-oriented processing model: Blocks are kept in GPU RAM until they are completely processed. This processing model is also known as vectorized processing [54]. Thus, the PCIe bottleneck can be further reduced by overlapping data transfers with computation. For certain queries, Yuan and others observed speedups up to 2.5x compared to no overlapping of processing and data transfers.

GPUDB compiles queries to *driver programs*. A driver program executes a query by calling pre-implemented GPU operators. Hence, GPUDB executes all queries on the GPU and the CPU performs only dispatcher and post processing tasks (i.e., the CPU is used less than 10 % of the time during processing SSB queries [62]).

---

[6] Source code available at: https://code.google.com/p/gpudb/.

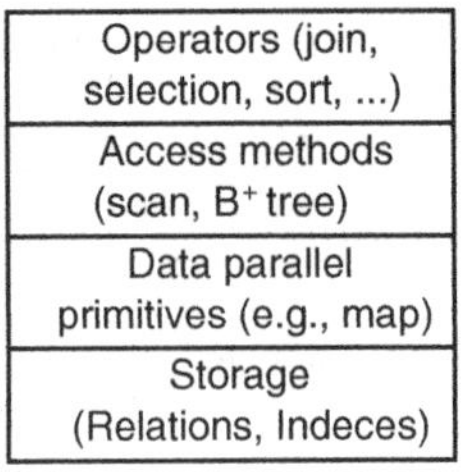

Operators (join, selection, sort, ...)
Access methods (scan, B⁺ tree)
Data parallel primitives (e.g., map)
Storage (Relations, Indeces)

Fig. 6. Execution engine of GPUQP, taken from [29]

Query Placement & Optimization: GPUDB has no support for executing queries on the CPU and GPU in parallel.

Transactions: Not supported.

GPUQP

He and others developed GPUQP[7], a relational query processing system, which stores data in-memory and uses the GPU to accelerate query processing [29]. In GPUQP, each relational operator can be executed on the CPU or the GPU (Fig. 6).

Storage System: GPUQP supports in-memory and disk-based processing. Apparently, GPUQP also attempts to keep data cached in GPU memory. Unfortunately, the authors do not provide any details about the used data placement strategy.

Storage Model: Furthermore, GPUQP makes use of columnar storage and query processing, which fits the hardware capabilities of modern CPUs and GPUs.

Processing Model: GPUQP's basic processing strategy is operator-at-a-time bulk processing. However, GPUQP is also capable of partitioning data for one operator and execute the operator on the CPU and the GPU concurrently. Nevertheless, the impact on the overall performance is small [29].

Query Placement & Optimization: GPUQP combines a Selinger-style optimizer [58] with an analytical cost model to select the cheapest query plan. For each operator, GPUQP allocates either the CPU, the GPU, or both processors (partitioned execution). The query optimizer splits a query plan to multiple sub-plans containing at most ten operators. For each sub-query, all possible plans are created and the cheapest sub-plan is selected. Finally, GPUQP combines the sub-plans to a final physical query plan.

---

[7] Source code available at: http://www.cse.ust.hk/gpuqp/.

He and others focus on optimizing single queries and do not discuss multi-query optimization. Furthermore, load-aware query scheduling is not considered and there is no discussion of scenarios with multiple GPUs.

Transactions: Not supported.

GPUTx

In order to investigate relational transaction processing on graphics cards, He and others developed GPUTx, a transaction processing engine that runs on the GPU [31].

Storage System & Model: GPUTx keeps all OLTP data inside the GPU's memory to minimize the impact of the PCIe bottleneck. It also applies a columnar data layout to fit the characteristics of modern GPUs.

Processing Model: The processing model is not built on relational operators as in GPUQP. Instead, GPUTx executes pre-compiled stored procedures, which are grouped into one GPU kernel. Incoming transactions are grouped in *bulks*, which are sets of transactions that are executed in parallel on the GPU.

Query Placement & Optimization: Since GPUTx performs the complete data processing on the GPU, query placement approaches are not needed.

Transactions: GPUTx is the only system in our survey – and that we are aware of – that supports running transactions on a GPU. It implements three basic transaction protocols: Two-phase locking, partition-based execution and k-set-based execution. The major finding of GPUTx is that locking-based protocols do not work well on GPUs. Instead, lock-free protocols such as partition-based execution or k-set should be used.

MapD

Mostak develops MapD, which is a data processing and visualization engine, combining traditional query processing capabilities of DBMSs with advanced analytic and visualization functionality [43]. One application scenario is the visualization of twitter messages on a road map[8], in which the geographical position of tweets is shown and visualized as heat map.

Storage System: The data processing component of MapD is a relational DBMS, which can handle data volumes that do not fit the main memory. MapD also tries to keep as much data in-memory as possible to avoid disk accesses.

---

[8] http://mapd.csail.mit.edu/tweetmap/.

Storage Model: MapD stores data in a columnar layout, and further partitions columns into *chunks*. A chunk is the basic unit of MapD's memory manager. The basic processing model of MapD is processing one operator-at-a-time. Due to the partitioning of data into chunks, it is also possible to process on a per-chunk basis. Hence, MapD is capable of applying block-oriented processing.

Processing Model: MapD processes queries by compiling a query to executable code for the CPU and GPU.

Query Placement & Optimization: The optimizer tries to split a query plan in parts, and processes each part on the most suitable processing device (e.g., text search using an index on the CPU and table scans on the GPU). MapD does not assume that an input data set fits in GPU RAM, and it applies a streaming mechanism for data processing.

Transactions: Not supported.

Ocelot

Heimel and others develop Ocelot[9], which is an OpenCL extension of MonetDB, enabling operator execution on any OpenCL capable device, including CPUs and GPUs [34] (Fig. 7).

Storage System: Ocelot's storage system is built on top of the in-memory model of MonetDB. Input data is automatically transferred from MonetDB to the GPU when needed by an operator. In order to avoid expensive transfers, operator results are typically kept on the GPU. They are only returned at the end of a query, or if the device memory is too filled to fulfill requests. Additionally, Ocelot implements a device cache to keep relevant input data available on the GPU.

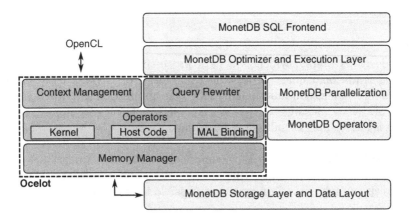

Fig. 7. The architecture of Ocelot, taken from [34]

---

[9] Source code available at: http://goo.gl/GHeUv.

Storage Model: Ocelot/MonetDB stores data column-wise in *Binary Association Tables* (BATs). Each BAT consists of two columns: One (optional) head storing object identifiers, and one (mandatory) tail storing the actual values.

Processing Model: Ocelot inherits the operator-at-a-time bulk processing model of MonetDB, but extends it by introducing lazy evaluation and making heavy use of the OpenCL event model to forward operator dependency information to the GPU. This allows the OpenCL driver to automatically interleave and reorder operations, e.g., to hide transfer latencies by overlapping the transfer with the execution of a previous operator.

Query Placement & Optimization: In MonetDB, each query plan is represented in the *MonetDB Assembly Language* (MAL) [35]. Ocelot reuses this infrastructures and adds a new query optimizer, which rewrites MAL plans by replacing data processing MAL instructions of vanilla MonetDB with the highly parallel OpenCL MAL instructions of Ocelot.

Query Placement & Optimization: Ocelot does not support cross-device processing, meaning it executes the complete workload either on the CPU or on the GPU.

Transactions: Not supported.

OmniDB

Zhang and others developed OmniDB[10], a GDBMS aiming for good code maintainability while exploiting all hardware resources for query processing [63]. The basic idea is to create a hardware oblivious database kernel (qkernel), which accesses the hardware via *adaptors*. Each adapter implements a common set of operators decoupling the hardware from the database kernel (Fig. 8).

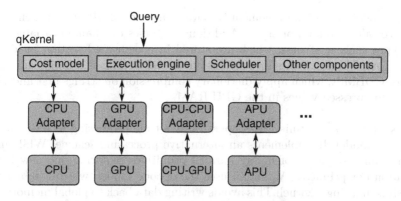

Fig. 8. OmniDB: Kernel adapter design, taken from [63]

---

[10] Source code available at: https://code.google.com/p/omnidb-paralleldbonapu/.

Storage System & Model: OmniDB is based on GPUQP, and hence, has similar architectural properties to GPUQP. OmniDB keeps data in-memory in a column-oriented data layout.

Processing Model: OmniDB schedules and processes *work units*, which can vary in granularity (e.g., a work unit can be a query, an operator, or a chunk of tuples). Although it is not explicitly mentioned in the paper [63], the fact that OmniDB can process also chunks of tuples is a strong indicator that it supports block-oriented processing.

Query Placement & Optimization: Regarding query placement and optimization, OmniDB chooses the processing device with highest throughput for a *work unit*. To avoid overloading a single device, OmniDB's scheduler ensures that the workload on one processing device may not exceed a certain percentage of the average workload on all processing devices. The cost model relies on the adapters to provide cost functions for the underlying processing devices.

Transactions: Not supported.

Virginian

Bakkum and others develop Virginian[11], which is a GPU-accelerated DBMS keeping data in main memory and supporting filter and aggregation operations on all processing devices [6].

Storage System: Virginian uses no traditional caching of operators, but *uniform virtual addressing* (UVA). This technique allows a GPU kernel to directly access data stored in pinned host memory. The accessed data is transferred over the bus transparently to the device and efficiently overlaps computation and data transfers.

Storage Model: Virgnian implements a data structure called *tablet*, which stores fixed size values column oriented. Additionally, tables can handle variable sized data types such as strings, which are stored in a dedicated section inside the tablet. Thus, Virginian supports strings on the GPU. This is a major difference to other GDBMSs, which apply dictionary compression on strings first and work only on compressed values in the GPU RAM.

Processing Model: Virginian uses operator-at-a-time processing as basic query-processing model. It implements an alternative processing scheme. While most systems call a sequence of highly parallel primitives requiring one new kernel invocation per primitive, Virginian uses the opcode model, which combines all primitives in a single kernel. This avoids writing data back to global memory and reading it again in the next kernel ultimately resulting in block-wise processing on the GPU.

---

[11] Source code available at: https://github.com/bakks/virginian.

Query Placement & Optimization: Virginian can either process queries on the CPU or on the GPU. Thus, there is no mechanism splitting up the workload between CPU and GPU processing devices and hence, no hybrid query optimizer is available.

Transactions: Not supported.

4.3 Classification

We now classify the surveyed systems according to the architectural properties discussed in Sect. 3.

Storage System: For all eight systems, it is clearly visible that they are designed with main-memory databases in mind, keeping a large fraction of the database in the CPU's main memory (Table 1). GPUQP and MapD also support disk-based data. However, since fetching data from disk is very expensive compared to transferring data over the PCIe bus [29], MapD and GPUQP also keep as much data as possible in main memory. Therefore, we mark all systems as main-memory storage and GPUQP and MapD additionally as disk-based storage.

Storage Model: All systems store their data in a columnar layout, there is no system using row-oriented storage (Table 1). One exception is Virginian, which stores data mainly column-oriented, but also kepps complete rows inside a tablet data structure. This representation is similar to PAX, which stores rows on one page, but stores all records column-wise inside a page [3].

Processing Model: The processing model varies between the surveyed systems (Table 2). The first observation is that no system uses a traditional tuple-at-a-time volcano model [26], as was hypothesized in Sect. 3. Most systems support

Table 1. Classification of Storage System and Storage Model – *Legend:* ✓ – Supported, × – Not Supported, ○ – Not Applicable

DBMS	Storage System		Storage Model	
	Main-Memory Storage	Disk-based Storage	Column Store	Row Store
CoGaDB	✓	×	✓	×
GPUDB	✓	×	✓	×
GPUQP	✓	✓	✓	×
GPUTx	✓	×	✓	×
MapD	✓	✓	✓	×
Ocelot	✓	×	✓	×
OmniDB	✓	×	✓	×
Virginian	✓	×	✓	×

operator-at-a-time bulk processing [42]. The only exception is GPUTx, which does not support OLAP workloads, because it is an optimized OLTP engine. Hence, we mark the processing model for GPUTx as not applicable. GPUDB, MapD, OmniDB, and Virginian have basic capabilities for block-oriented processing. Additionally, GPUDB and MapD apply a compilation-based query processing strategy.[12] Virginian does not support query compilation. Instead, it uses a single GPU kernel that implements a virtual machine, which calls other GPU kernels (the primitives) in the context of the same kernel, efficiently saving the overhead of reading and writing the result from the GPU's main memory.

Query Placement and Optimization: We identify two major groups of systems: The first group performs nearly all data processing on one processing device (GPUDB, GPUTx, Ocelot, Virginian), whereas the second group is capable of splitting the workload in parts, which are then processed in parallel on the CPU and the GPU (CoGaDB, GPUQP, MapD, OmniDB) (Table 3). We mark systems in the first group as systems that support only single-device processing (SDP), whereas systems of the second group are capable of using multiple devices and thereby allowing cross-device processing (CDP). Note that a system supporting CDP is also capable of executing the complete workload on one processing device (SDP). The hybrid query optimization approaches of CoGaDB, GPUQP, MapD, and OmniDB are mostly greedy strategies or other simple heuristics. It is still an open question how to efficiently trade off between inter-processor parallelization and costly data transfers to achieve optimal performance. For instance: So far, there are no query optimization approaches for machines having multiple GPUs.

Table 2. Classification of Processing Model – *Legend:* ✓ – Supported, × – Not Supported, ○ – Not Applicable

DBMS	Processing Model		
	Operator-at-a-Time	Block-at-a-Time	Just-in-Time Compilation
CoGaDB	✓	×	×
GPUDB	✓	✓	✓
GPUQP	✓	×	×
GPUTx	○	○	○
MapD	✓	✓	✓
Ocelot	✓	×	×
OmniDB	✓	✓	×
Virginian	✓	✓	×

---

[12] Note that both systems still apply a block-oriented processing model. This is due to the nature of compilation-based strategies, as discussed in Sect. 3.

Table 3. Classification of Query Processing – *Legend:* ✓ – Supported, × – Not Supported, ○ – Not Applicable

DBMS	Query Processing	
	Single-Device Processing	Cross-Device Processing
CoGaDB	✓	✓
GPUDB	✓	×
GPUQP	✓	✓
GPUTx	✓	×
MapD	✓	✓
Ocelot	✓	×
OmniDB	✓	✓
Virginian	✓	×

Transaction Processing: Apart from GPUTx, none of the surveyed GDBMSs support transactions (Table 4). GPUTx keeps data strictly in the GPU's RAM, and needs to transfer only incoming transactions to the GPU and the result back to the CPU. Since GPUTx achieved a 4–10 times higher throughput than a comparable CPU-based OLTP engine, there is a need for further research in the area of transaction processing in GDBMSs so that OLTP systems can also benefit from GPU acceleration. Apparently, online analytical processing and online transactional processing can be significantly accelerated by using GPU acceleration. However, it is not yet clear which workload type is more suitable for which processing device type. Furthermore, the efficient combination of OLTP/OLAP workloads is still an active research field (e.g., Kemper and Neumann [38]). Thus, it is an open question whether and under which circumstances GPU-acceleration is beneficial for such hybrid OLTP/OLAP workloads.

Portability: The only GDBMSs having a portable, hardware-oblivious database architecture are Ocelot and OmniDB. All other systems are either tailored to a vendor specific programming framework or have no technique to hide the details of the device-specific operators in the architecture. Ocelot's approach has the advantage that only a single set of parallel database operators has to be implemented, which can then be mapped to all processing devices supporting OpenCL (e.g., CPUs, GPUs, or Xeon Phis). By contrast, OmniDB uses an adapter interface, in which each adapter provides a set of operators and cost functions for a certain processing-device type. It is unclear, which approach will lead to the best performance/maintainability ratio, and how large the performance loss is compared to a hardware-aware system. However, if portability can be achieved with only a small performance degradation, it would substantially benefit the maintainability and applicability of GDBMSs [63]. Hence, the trend towards hardware-oblivious DBMSs is likely to continue.

Table 4. Classification of Transaction Support and Portability – *Legend:* ✓ – Supported, × – Not Supported, ○ – Not Applicable

DBMS	Transaction Support	Portability	
		Hardware Aware	Hardware Oblivious
CoGaDB	×	✓	×
GPUDB	×	✓	×
GPUQP	×	✓	×
GPUTx	✓	✓	×
MapD	×	✓	×
Ocelot	×	×	✓
OmniDB	×	×	✓
Virginian	×	✓	×

4.4   Potential Optimizations for GDBMSs

We will now discuss and summarize potential optimizations, which a GDBMS may implement to make full use of the underlying hardware in a hybrid CPU/GPU system. Additionally, we briefly discuss existing approaches for each optimization. As already discussed, data transfers have the highest impact on GDBMS performance. Hence, every optimization avoiding or minimizing the impact of data transfers are mandatory. We refer to these optimizations as cross-device optimizations. Based on our surveyed systems, we could identify the following *cross-device optimizations*:

Efficient Data Placement Strategy: There are two possibilities to manage the GPU RAM. The first possibility is an explicit management of data on GPUs using a buffer-management algorithm. The second possibility is using mechanisms such as *Unified Virtual Addressing* (UVA), which enables a GPU kernel to directly access the main memory. Kaldewey and others observed a significant performance gain (3-8x) using UVA for Hash Joins on the GPU compared to the CPU [37]. Furthermore, data has not to be kept consistent between CPU and GPU, because there is no "real" copy in the GPU RAM. However, this advantage can also be a disadvantage, because caching data in the GPU RAM can avoid the data transfer from the CPU to the GPU.

GPU-aware Query Optimizer: A GDBMS should make use of all processing devices to maximize performance. Therefore, it should offload operations to the GPU. However, offloading single operations of a query plan does not necessarily accelerate performance. Hence, a GPU-aware optimizer has to identify sub plans of a query plan, which it can process on the CPU or the GPU [29]. Furthermore, the resulting plan should minimize the number of copy operations [15]. Since optimizers are typically cost based, a GDBMS needs for each GPU operator a cost model. The most common approach is to use analytical models (e.g., Manegold and others for the CPU [40] and He and others for the GPU [29]). However, with the increasing hardware complexity, machine-learning-based models become increasingly popular [14].

Data Compression: The data placement and query optimization techniques attempt to avoid data transfers as much as possible. To reduce overhead in case a GDBMS has to perform data transfers, the data volume can be reduced by compression techniques. Thus, compression can significantly decrease processing costs [62]. Fang and others discussed an approach, which combines different lightweight compression techniques to compress data at the GPU [23]. They developed a planner for cascading compression techniques, which decides on a suitable subset and order of available compression techniques. Przymus and Kaczmarski focused on compression for time-series databases on the GPU [50]. Andrzejewski and Wrembel discussed compression of bitmap indexes on the GPU [4].

Overlap of Data Transfer and Processing: The second way to accelerate processing, in case a data transfer needs to performed, is overlapping the execution of a GPU operator with a data transfer operation [6,62]. This optimization keeps all hardware components busy, and basically narrows down the performance of the system to the PCIe bus bandwidth.

Pinned Host Memory: The third way to accelerate query processing in case we have to perform a copy operation is keeping data in pinned host memory. This optimization saves one indirection, because the DMA controller can transmit data directly to the device [62]. Otherwise, data has to be copied in pinned memory first, introducing additional latency in data transmission. However, using pinned host memory has the drawback that the amount of available pinned host memory is much smaller than the amount of unpinned memory (i.e., memory that can be paged to disk by the virtual memory manager) [56]. Therefore, a GDBMS has to decide which data it should keep in pinned host memory. It is still an open issue how much memory should be spent on a pinned host memory buffer for faster data transfers to the GPU.

Figure 9 illustrates the identified cross-device optimizations and the relationships between them.

The second class of optimizations we identified, targets the efficiency of operator execution on a single processing device. We refer to this class of optimizations as *device-dependent optimizations*. Since we focus on GPU-aware systems, we only discuss optimizations for GPUs. Based on the surveyed systems, we summarize the following GPU-dependent optimizations:

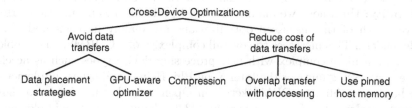

Fig. 9. Cross-device optimizations

Block-oriented Query Processing: A GDBMS can avoid the overhead of writing results of an operator back to a processing device's main memory by processing data on a per block basis rather than on a per operator basis. The idea is to process data already stored in the cache (CPU) or shared memory (GPU), which saves memory bandwidth and significantly increases performance of query processing [11,62]. Additionally, block-oriented processing is a necessary prerequisite for overlapping processing and data transfer for single operations and allows for a more fine grained workload distribution on available processing devices [63]. Note that traditional pipelining of blocks between GPU operators is not possible, because inter-kernel communication is undefined [15]. While launching a new kernel for each block is likely to be expensive, query compilation and kernel fusion are promising ways to allow block-oriented processing on the GPU as well.

Compilation-based Query Processing: Compiling queries to executable code is a common optimization in main-memory DBMSs [20,44,60]. As already discussed, query compilation allows for block-oriented processing on GPUs as well and achieves a significant speedup compared to primitive-based query processing (e.g., operator-at-a-time processing [29]). However, query compilation introduces additional overhead, because compiling a query to executable code typically is more expensive than building a physical query execution plan. Yuan and others overcome this shortcoming by pre-compiling operators. Thus, they only need to compile the query plan itself to a driver program [62]. A similar approach called *kernel weaver* is used by Wu and others [61]. They combine CUDA kernels for relational primitives into one kernel. This has the advantage that the optimization scope is larger and the compiler can perform more optimizations. However, the disadvantage is the increased compilation time. Rauhe and others introduce in their approach two processing phases: compute and accumulate. In the compute phase, a number of threads are assigned to a partition of the input data and each thread performs all operations of a query on one tuple and then, continues with the next tuple, until the thread processed its partition. In the accumulate phase, the intermediate results are combined to the final result [53].

All-in-one Kernel: A promising alternative to compilation-based approaches is to combine all relational primitives in one kernel [6]. Thus, a relational query has to be translated to a sequence of op codes. An op code identifies the next primitive to be executed. Therefore, it is basically an on-GPU virtual machine, which saves the initial overhead of query compilation. However, the drawback is a limited optimization scope compared to kernel weaver [61].

Portability: Until now, we mainly discussed performance optimizations. However, each of the discussed optimizations are mainly implemented device dependent. This increases the overall complexity of a GDBMS. The problem gets even more complex with new processing device types such as accelerated processing units or the Intel Xeon Phi. Heimel and others implemented a hardware oblivious DBMS kernel in OpenCL and still achieved a significant acceleration of query processing [34]. Zhang and others implemented *q-kernel*, a hardware-oblivious database kernel using device adapters to the

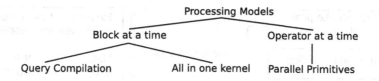

Fig. 10. Device-dependent optimizations: Efficient processing models

underlying processing devices [63]. It is still not clear which part of a kernel should be hardware oblivious and which part should be hardware aware. For the parts that have to be hardware aware, modern software engineering methods such as software product lines can be used to manage the GDBMS's complexity [19].

Figure 10 illustrates the identified device-dependent optimizations and the relationships between them.

4.5   A Reference Architecture for GPU-Accelerated DBMSs

Based on our in-depth survey of existing GDBMSs, we now derive a reference architecture for GDBMSs. After careful consideration of all surveyed systems, we decided to use the GPUQP [29]/OmniDB [63] architecture as basis for our reference architecture, because they already include a major part of the common properties of the surveyed systems. We illustrate the reference architecture in Fig. 11.

We will describe the query-evaluation process in a top-down view. On the upper levels of the query stack, a GPU-accelerated DMBS is virtually identical to a "traditional" DBMS. It includes functionality for integrity control, parsing SQL queries, and performing logical optimizations on queries. Major differences between main-memory DBMSs and GDBMSs emerge in the physical optimizer. While classical systems choose the most suitable access structure and algorithm to operate on the access structure, a GPU-accelerated DBMS has to additionally decide for each operator on a processing device. For this task, a GDBMS needs refined[13] cost models that also predict the cost for GPU and CPU operations. Based on these estimates, a scheduler can allocate the cheapest processing device. Furthermore, a query should make use of multiple processing devices to speed up execution. Hence, the physical optimizer has to optimize hybrid CPU/GPU query plans, which significantly increases the optimization space.

Relational operations are implemented in the next layer. These operators typically use access structures to process data. In GDBMSs, access structures have to be reimplemented on GPUs to achieve a high efficiency. However, depending

---

[13] Since these models need to be able to estimate comparable operator runtimes across different devices, we and others [13] argue that dynamic cost models, which apply techniques from Machine Learning to adapt to the current hardware, are likely required here.

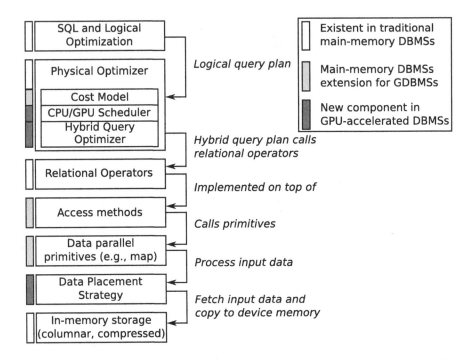

Fig. 11. Layered architecture of GDBMSs

on the processing device chosen by the CPU/GPU scheduler, different access structures are available. This is an additional dependency the query optimizer needs to take into account.

Then, a set of parallel primitives can be applied to an access structure to process a query. In this component, the massive parallelism of CPUs and GPUs is fully used to speed up query processing. However, a GPU operator can only work on data stored in GPU memory. Hence, all access structures are built on top of a data-placement component, that caches data on a certain processing device, depending on the access patterns of the workload (e.g., certain columns for column scans or certain nodes of tree indexes [8,57]). Note that the data-placement strategy is the most performance critical component in a GDBMS due to the major performance impact of data transfers.

The backbone of a GDBMS is a typical in-memory storage, which frequently stores data in a column-oriented format.[14] Compression techniques are not only beneficial in keeping the major part of a database in-memory, compression also reduces the impact of the PCIe bottleneck.

---

[14] We are aware that some in-memory DBMSs can also store data row-oriented, such as HyPer [38]. However, in GDBMSs, row-oriented storage either increases the data volume to be transferred or requires a projection operation before the transfer. A row-oriented layout also makes it difficult to achieve optimal memory access patterns on a GPU.

4.6 Summary: Extension Points for Main-Memory DBMSs

In summary, we can extend most main-memory DBMSs supporting column-oriented data layout and bulk processing to be GPU-accelerated DBMSs. We identify the following extension points: Cost models, CPU/GPU scheduler, hybrid query optimizer, access structures and algorithms for the GPU, and a data placement strategy.

Cost Models: For each processor, we need to estimate the execution time of an operator. This can be either done by analytical cost models (e.g., Manegold and others for CPUs [40] and He and others for GPUs [29]) or learning-based approaches (e.g., Breß and others [14] or Ilić and Sousa [36]).

CPU/GPU Scheduler: Based on the cost models, a scheduler needs to allocate processing devices for a set of operators (e.g., CHPS from Ilić and Sousa, HyPE from Breß and others [14], or StarPU from Augonnet and others [5]).

Hybrid Query Optimizer: The query optimizer needs to consider the data transfer bottleneck and memory requirements of operators to create a suitable physical execution plan. Thus, the optimizer should make use of cost models, a CPU/GPU scheduler, and heuristics minimizing the time penalty of data transfers (e.g., HyPE from Breß and others [14]).

Access structures and algorithms for the GPU: In order to support GPU-acceleration, a DBMS needs to implement access structures on the GPU (e.g., columns or B$^+$-trees) and operators that work on them. Here, the most approaches were developed [7,21,29,32,48,49].

Data Placement Strategy: A DBMS needs to keep track of which data is stored on the GPU, and which access structure needs to be transferred to GPU memory [29]. Aside from a manual memory management, it is also possible to use techniques such as UVA and let the GPU driver handle the data transfers transparently to the DBMS [62]. However, this may result in less efficiency because a manual memory management can exploit knowledge about the DBMS and the workload.

Implementing these extensions is a necessary precondition for a DBMS to support GPU co-processing efficiently.

5 Open Challenges and Research Questions

In this section, we identify *open challenges* for GPU-accelerated DBMSs. We differentiate between two major classes of challenges, namely the IO bottleneck, which includes disk IO as well as data transfers between CPU and GPU, and query optimization.

5.1 IO Bottleneck

In a GDBMS, there are two major IO bottlenecks. The first is the classical disk IO, and the second bottleneck is the PCIe bus. As for the latter bottleneck, we can differ between avoiding and reducing the impact of the bottleneck.

Disk-IO Bottleneck: GPU-accelerated operators are of little use for disk-based database systems, where most time is spent on disk I/O. Since the GPU improves performance only once the data is in main memory, time savings will be small compared to the total query runtime. Furthermore, disk-resident databases are typically very large, making it harder to find an optimal data placement strategy. However, database systems can benefit from GPUs even in scenarios where not the complete database fits into main memory. As long as the *hot data* fits into main memory, GPUs can accelerate data processing. It is still an open problem to which degree a database has to fit into the CPU's main memory, so GPU acceleration pays off.

Data Placement Strategy: GPU-accelerated databases try to keep relational data cached on the device to avoid data transfer. Since device memory is limited, this is often only possible for a subset of the data. Deciding which part of the data should be offloaded to the GPU – finding a *data placement strategy* – is a difficult problem that currently remains unsolved.

Reducing PCIe Bus Bottleneck: Data transfers can be significantly accelerated by keeping data in pinned host memory. However, the amount of available pinned memory is much more limited compared to the amount of available virtual memory. Therefore, a GDBMS has to decide which data to keep in pinned memory. Since data is typically cached in GPU memory, a GDBMS needs a multi-level caching technique, which is yet to be found.

5.2 Query Optimization

In GDBMSs, query processing and optimization have to cope with new challenges. We identify as major open challenges a generic cost model, an increased complexity of query optimization due to the larger optimization space, insufficient support for using multi-processing devices for query-compilation approaches, and accelerating different workload types.

Generic Cost Model: From the query-optimization perspective, a GDBMS needs a cost model to perform cost-based optimization. In this area, two basic cost-model classes have emerged. The first class consists of analytical cost models and the second class makes use of machine-learning approaches to learn cost models for some training data. While analytical cost models excel in computational efficiency, learning-based strategies need no knowledge about the underlying hardware and can adapt to changing data. It is still open which kind of cost model is optimal for GDBMSs.

Increased Complexity of Query Optimization: Having the option of running operations on a GPU increases the complexity of query optimization: The plan search space is significantly larger and a cost function that compares run-times across architectures is required. While there has been prior work in this direction [14, 15, 29], GPU-aware query optimization remains an open challenge.

Query Compilation for Multiple Devices: With the upcoming trend of query compilation, the basic problem of processing-device allocation remains the same as in traditional query optimization. However, as of now, the available compilation approaches only compile complete queries for either the CPU or the GPU. It is still an open challenge how to compile queries to code that uses more than one processing device concurrently.

Considering different Workload Types: OLTP as well as OLAP workloads can be significantly accelerated using GPUs. Furthermore, it became common to have a mix of both workload types in a single system. It remains open, which workload types are more suited for which processing-device type and how to efficiently schedule OLTP and OLAP queries on the CPU and the GPU.

6   Conclusion and Future Directions

The performance of modern processors is no longer bound primarily by transistor density but by a fixed energy budget, the *power wall* [12]. Whereas CPUs often spend additional chip space on more cache capacity, other processors spend most of their chip space on light-weight cores, which omit heavy control logic and are thus, more energy efficient. Therefore, future machines will likely consist of a set of heterogeneous processors, having CPUs and specialized co-processors such as GPUs, Multiple Integrated Cores (MICs), or FPGAs. Hence, the question of using co-processors in databases is not *why* but *how* we can do this most efficiently.

The pioneer of modern co-processors is the GPU, and many prototypes of GPU-accelerated DBMSs have emerged over the past seven years implementing new co-processing approaches and proposing new system architectures. We argue that we need to take into account tomorrows hardware in today's design decisions. Therefore, in this paper, we theoretically explored the design space of GPU-aware database systems. In summary, we argue that a GDBMS should be an in-memory, column-oriented DBMS using the block-at-a-time processing model, possibly extended by a just-in-time-compilation component. The system should have a query optimizer that is aware of co-processors and data-locality, and is able to distribute a workload across all available (co-)processors.

We validated these findings by surveying the implementation details of eight existing GDBMSs and classifying them along the mentioned dimensions. Additionally, we summarized common optimizations implemented in GDBMSs and inferred a reference architecture for GDBMSs, which may act as a starting point in integrating GPU-acceleration in popular main-memory DBMSs. Finally, we identified potential *open challenges* for further development of GDBMSs.

Our results are not limited to GPUs, but should also be applicable to other co-processors. The existing techniques can be applied to virtually all massively parallel processors having dedicated high-bandwidth memory with limited storage capacity.

Acknowledgements. We thank Tobias Lauer from Jedox AG and the anonymous reviewers of the GPUs in Databases Workshop for their helpful feedback on the workshop version of this paper [17]. We thank Jens Teubner from TU Dortmund University, Michael Saecker from ParStream GmbH, and the anonymous reviewers of the TLDKS journal for their helpful comments on the journal version of this paper.

References

1. Palo GPU accelerator. White Paper (2010)
2. Parstream - turning data into knowledge. White Paper, November 2010
3. Ailamaki, A., DeWitt, D.J., Hill, M.D., Skounakis, M.: Weaving relations for cache performance. In: VLDB, pp. 169–180. Morgan Kaufmann Publishers Inc. (2001)
4. Andrzejewski, W., Wrembel, R.: GPU-WAH: applying GPUs to compressing bitmap indexes with word aligned hybrid. In: Bringas, P.G., Hameurlain, A., Quirchmayr, G. (eds.) DEXA 2010, Part II. LNCS, vol. 6262, pp. 315–329. Springer, Heidelberg (2010)
5. Augonnet, C., Thibault, S., Namyst, R., Wacrenier, P.-A.: StarPU: a unified platform for task scheduling on heterogeneous multicore architectures. Concurr. Comput. Pract. Exp. **23**(2), 187–198 (2011)
6. Bakkum, P., Chakradhar, S.: Efficient data management for GPU databases (2012). http://pbbakkum.com/virginian/paper.pdf
7. Bakkum, P., Skadron, K.: Accelerating SQL database operations on a GPU with CUDA. In: GPGPU, pp. 94–103. ACM (2010)
8. Beier, F., Kilias, T., Sattler, K.-U.: GiST scan acceleration using coprocessors. In: DaMoN, pp. 63–69. ACM (2012)
9. Binnig, C., Hildenbrand, S., Färber, F.: Dictionary-based order-preserving string compression for main memory column stores. In: SIGMOD, pp. 283–296. ACM (2009)
10. Boncz, P.A., Kersten, M.L., Manegold, S.: Breaking the memory wall in MonetDB. Commun. ACM **51**(12), 77–85 (2008)
11. Boncz, P.A., Zukowski, M., Nes, N.: MonetDB/X100: hyper-pipelining query execution. In: CIDR, pp. 225–237 (2005)
12. Borkar, S., Chien, A.A.: The future of microprocessors. Commun. ACM **54**(5), 67–77 (2011)
13. Breß, S.: Why it is time for a HyPE: a hybrid query processing engine for efficient GPU coprocessing in dbms. The VLDB PhD Workshop, PVLDB **6**(12), 1398–1403 (2013)
14. Breß, S., Beier, F., Rauhe, H., Sattler, K.-U., Schallehn, E., Saake, G.: Efficient co-processor utilization in database query processing. Inf. Syst. **38**(8), 1084–1096 (2013)
15. Breß, S., Geist, I., Schallehn, E., Mory, M., Saake, G.: A framework for cost based optimization of hybrid CPU/GPU query plans in database systems. Control Cybern. **41**(4), 715–742 (2012)
16. Breß, S., Haberkorn, R., Ladewig, S.: CoGaDB reference manual (2014). http://wwwiti.cs.uni-magdeburg.de/iti_db/research/gpu/cogadb/0.3/doc/refman.pdf
17. Breß, S., Heimel, M., Siegmund, N., Bellatreche, L., Saake, G.: Exploring the design space of a GPU-aware database architecture. In: Catania, B., et al. (eds.) New Trends in Databases and Information Systems. AISC, vol. 241, pp. 225–234. Springer, Heidelberg (2014)

18. Breß, S., Siegmund, N., Bellatreche, L., Saake, G.: An operator-stream-based scheduling engine for effective GPU coprocessing. In: Catania, B., Guerrini, G., Pokorný, J. (eds.) ADBIS 2013. LNCS, vol. 8133, pp. 288–301. Springer, Heidelberg (2013)

19. Broneske, D., Breß, S., Heimel, M., Saake, G.: Toward hardware-sensitive database operations. In: EDBT, pp. 229–234. OpenProceedings.org (2014)

20. Dees, J., Sanders, P.: Efficient many-core query execution in main memory column-stores. In: ICDE, pp. 350–361. IEEE (2013)

21. Diamos, G., Wu, H., Lele, A., Wang, J., Yalamanchili, S.: Efficient relational alge-bra algorithms and data structures for GPU. Technical report, Center for Experi-mental Research in Computer Systems (CERS) (2012)

22. Fang, R., He, B., Lu, M., Yang, K., Govindaraju, N.K., Luo, Q., Sander, P.V.: GPUQP: query co-processing using graphics processors. In: SIGMOD, pp. 1061–1063. ACM (2007)

23. Fang, W., He, B., Luo, Q.: Database compression on graphics processors. PVLDB **3**, 670–680 (2010)

24. Gaster, B.R., Howes, L., Kaeli, D., Mistry, P., Schaa, D.: Heterogeneous Computing With Opencl. Elsevier Sci. Technol. 1–2 (2012)

25. Ghodsnia, P.: An in-GPU-memory column-oriented database for processing ana-lytical workloads. In: The VLDB PhD Workshop. VLDB Endowment (2012)

26. Graefe, G.: Encapsulation of parallelism in the volcano query processing system. In: SIGMOD, pp. 102–111. ACM (1990)

27. Gregg, C., Hazelwood, K.: Where is the data? why you cannot debate CPU vs. GPU performance without the answer. In: ISPASS, pp. 134–144. IEEE (2011)

28. He, B., Fang, W., Luo, Q., Govindaraju, N.K., Wang, T.: Mars: a mapreduce framework on graphics processors. In: PACT, pp. 260–269. ACM (2008)

29. He, B., Lu, M., Yang, K., Fang, R., Govindaraju, N.K., Luo, Q., Sander, P.V.: Relational query co-processing on graphics processors. In: ACM Transactions on Database System, vol. 34. ACM (2009)

30. He, B., Yang, K., Fang, R., Lu, M., Govindaraju, N., Luo, Q., Sander, P.: Relational joins on graphics processors. In: SIGMOD, pp. 511–524. ACM (2008)

31. He, B., Yu, J.X.: High-throughput transaction executions on graphics processors. PVLDB **4**(5), 314–325 (2011)

32. He, J., Lu, M., He, B.: Revisiting co-processing for hash joins on the coupled CPU-GPU architecture. PVLDB **6**(10), 889–900 (2013)

33. Heimel, M., Markl, V.: A first step towards GPU-assisted query optimization. In: ADMS. VLDB Endowment (2012)

34. Heimel, M., Saecker, M., Pirk, H., Manegold, S., Markl, V.: Hardware-oblivious parallelism for in-memory column-stores. PVLDB **6**(9), 709–720 (2013)

35. Idreos, S., Groffen, F., Nes, N., Manegold, S., Mullender, K.S., Kersten, M.L.: MonetDB: Two decades of research in column-oriented database architectures. IEEE Data Eng. Bull. **35**(1), 40–45 (2012)

36. Ilić, A., Sousa, L.: CHPS: an environment for collaborative execution on heteroge-neous desktop systems. Int. J. Netw. Comput. **1**(1), 96–113 (2011)

37. Kaldewey, T., Lohman, G., Mueller, R., Volk, P.: GPU join processing revisited. In: DaMoN, pp. 55–62. ACM (2012)

38. Kemper, A., Neumann, T.: HyPer: a hybrid OLTP&OLAP main memory database system based on virtual memory snapshots. In: ICDE, pp. 195–206. IEEE (2011)

39. Kossmann, D.: The state of the art in distributed query processing. ACM Comput. Surv. **32**(4), 422–469 (2000)

40. Manegold, S., Boncz, P., Kersten, M.L.: Generic database cost models for hierarchical memory systems. In: PVLDB, pp. 191–202. VLDB Endowment (2002)

41. Manegold, S., Boncz, P.A., Kersten, M.L.: Optimizing database architecture for the new bottleneck: memory access. VLDB J. **9**(3), 231–246 (2000)

42. Manegold, S., Kersten, M.L., Boncz, P.: Database architecture evolution: mammals flourished long before dinosaurs became extinct. PVLDB **2**(2), 1648–1653 (2009)

43. Mostak, T.: An overview of MapD (massively parallel database). White Paper, Massachusetts Institute of Technology, April 2013. http://geops.csail.mit.edu/docs/mapd_overview.pdf

44. Neumann, T.: Efficiently compiling efficient query plans for modern hardware. PVLDB **4**(9), 539–550 (2011)

45. NVIDIA. NVIDIA CUDA C programming guide, pp. 31–36, 40, 213–216, Version 6.0. (2014). http://docs.nvidia.com/cuda/pdf/CUDA_C_Programming_Guide.pdf. Accessed 21 April 2014

46. Owens, J.D., Luebke, D., Govindaraju, N., Harris, M., Krger, J., Lefohn, A.E., Purcell, T.J.: A survey of general-purpose computation on graphics hardware. Comput. Graph. Forum **26**(1), 80–113 (2007)

47. Pirk, H.: Efficient cross-device query processing. In: The VLDB PhD Workshop. VLDB Endowment (2012)

48. Pirk, H., Manegold, S., Kersten, M.: Accelerating foreign-key joins using asymmetric memory channels. In: ADMS, pp. 585–597. VLDB Endowment (2011)

49. Pirk, H., Manegold, S., Kersten, M.: Waste not... efficient co-processing of relational data. In: ICDE. IEEE (2014)

50. Przymus, P., Kaczmarski, K.: Dynamic compression strategy for time series database using GPU. In: Catania, B., et al. (eds.) New Trends in Databases and Information Systems. AISC, vol. 241, pp. 235–244. Springer, Heidelberg (2014)

51. Przymus, P., Kaczmarski, K., Stencel, K.: A bi-objective optimization framework for heterogeneous CPU/GPU query plans. In: CS&P, pp. 342–354. CEUR-WS (2013)

52. Rabl, T., Poess, M., Jacobsen, H.-A., O'Neil, P., O'Neil, E.: Variations of the star schema benchmark to test the effects of data skew on query performance. In: ICPE, pp. 361–372. ACM (2013)

53. Rauhe, H., Dees, J., Sattler, K.-U., Faerber, F.: Multi-level parallel query execution framework for CPU and GPU. In: Catania, B., Guerrini, G., Pokorný, J. (eds.) ADBIS 2013. LNCS, vol. 8133, pp. 330–343. Springer, Heidelberg (2013)

54. Răducanu, B., Boncz, P., Zukowski, M.: Micro adaptivity in vectorwise. In: SIGMOD, pp. 1231–1242. ACM (2013)

55. Saecker, M., Markl, V.: Big data analytics on modern hardware architectures: a technology survey. In: Aufaure, M.-A., Zimányi, E. (eds.) eBISS 2012. LNBIP, vol. 138, pp. 125–149. Springer, Heidelberg (2013)

56. Sanders, J., Kandrot, E.: CUDA by Example: An Introduction to General-Purpose GPU Programming, 1st edn. Addison-Wesley Professional, Upper Saddle River (2010)

57. Schäler, M., Grebhahn, A., Schröter, R., Schulze, S., Köppen, V., Saake, G.: QuEval: beyond high-dimensional indexing à la carte. PVLDB **6**(14), 1654–1665 (2013)

58. Selinger, P.G., Astrahan, M.M., Chamberlin, D.D., Lorie, R.A., Price, T.G.: Access path selection in a relational database management system. In: SIGMOD, pp. 23–34. ACM (1979)

59. Tsirogiannis, D., Harizopoulos, S., Shah, M.A.: Analyzing the energy efficiency of a database server. In: SIGMOD, pp. 231–242. ACM (2010)

60. Viglas, S.D.: Just-in-time compilation for SQL query processing. PVLDB **6**(11), 1190–1191 (2013)
61. Wu, H., Diamos, G., Cadambi, S., Yalamanchili, S.: Kernel weaver: automatically fusing database primitives for efficient GPU computation. In: MICRO, pp. 107–118. IEEE (2012)
62. Yuan, Y., Lee, R., Zhang, X.: The yin and yang of processing data warehousing queries on GPU devices. PVLDB **6**(10), 817–828 (2013)
63. Zhang, S., He, J., He, B., OmniDB, M.L.: Towards portable and efficient query processing on parallel CPU/GPU architectures. PVLDB **6**(12), 1374–1377 (2013)
64. Zhong, J., He, B.: Medusa: simplified graph processing on gpus. IEEE Trans. Parallel Distrib. Syst. **99**, 1–14 (2013)
65. Zhong, J., He, B.: Parallel graph processing on graphics processors made easy. PVLDB **6**(12), 1270–1273 (2013)

Compression Planner for Time Series Database with GPU Support

Piotr Przymus[1]([⊠]) and Krzysztof Kaczmarski[2]

[1] Nicolaus Copernicus University, Toruń, Poland
eror@mat.umk.pl
[2] Warsaw University of Technology, Warsaw, Poland
k.kaczmarski@mini.pw.edu.pl

Abstract. Nowadays, we can observe increasing interest in processing and exploration of time series. Growing volumes of data and needs of efficient processing pushed research in new directions. This paper presents a lossless lightweight compression planner intended to be used in a time series database system. We propose a novel compression method which is ultra fast and tries to find the best possible compression ratio by composing several lightweight algorithms tuned dynamically for incoming data. The preliminary results are promising and open new horizons for data intensive monitoring and analytic systems.

Keywords: Time series database · Lightweight compression · Lossless compression · GPU · CUDA · GPGPU · Compression optimization

1 Introduction

Background – Time Series Databases. Specialized time series databases play important role in industry storing monitoring data for analytical purposes. These systems are expected to process and store millions of data points per minute, 24 h a day, seven days a week, reading terabytes of logs. Due to regression errors checking and early malfunction prediction these data must be kept with fine grained resolution including all details. Solutions like OpenTSDB [17], TempoDB [4] and others deal very well with these kind of tasks. Most of them work on a clone of Big Table approach from Google [8], a distributed hash table with mutual ability to write and read data in the same time.

Querying large volumes of time series may be time consuming and even in case of big clusters leads to system slowdown. On the other hand monitoring of any infrastructure requires real-time or near real-time response. What is more

P. Przymus: The project was partially funded by Marshall of Kuyavian-Pomeranian Voivodeship in Poland with the funds from European Social Fund (EFS) in the form of a PhD scholarships. "Krok w przyszłość – stypendia dla doktorantów V edycja" (Step in the future – PhD scholarships V edition).

K. Kaczmarski: The project was partially funded by National Science Centre, decision DEC-2012/07/D/ST6/02483.

© Springer-Verlag Berlin Heidelberg 2014
A. Hameurlain et al. (Eds.): TLDKS XV, LNCS 8920, pp. 36–63, 2014.
DOI: 10.1007/978-3-662-45761-0_2

important it is hard to predict a priori what kind of queries may be needed. Various problems may be only investigated by checking all possible correlations. Therefore a system must perform random queries on large data sets. Classical databases even using large computational clusters and map reduce approach can hardly fulfil this requirement since time series processing not only requires large volumes of data but also has computational demands: interpolation, integration and aggregation of millions of time series.

The above problems may be easily handled by a database system equipped with a GPU device used as a coprocessor [7]. An average internet service with about 10 thousands of simultaneously working users may generate around 80 GB of logs every day. If we consider an in-memory database system these data after a compression could fit into two NVIDIA Tesla devices and an average query may be processed within milliseconds compared to seconds or minutes in case of standard systems.

GPU device has its own memory or a separate area in the CPU main memory. CPU and GPU may only cooperate in a *shared nothing architecture.*Thus, the data has to be explicitly transferred from the CPU main memory to the GPU main memory and then back to CPU. Additional data transfer in the pipeline of a query processing often introduces significant overhead which cannot be mitigated. This cost is therefore an important component of the query execution time prediction.

Time Series Compression. Big Table based systems compress data before writing to a long-term storage. It is much more efficient to store data for some time in memory or in a disk buffer and compress it before flushing to disk. This process is known as a *table row rolling.* Systems like HBase [1], Casandra [10] and others offer compression for entire column family. This kind of general purpose compression is not optimized for particular data being stored (i.e. various time series with different compression potential stored in one column family). Similarly in-memory database systems based on GPU processing (like ParStream [3]) tend to pack as many data into GPU devices global memory as possible.

Compression not only improves overall system behaviour by optimizing data transfer but also enables GPU co-processing by minimization of additional data transfer costs. Figure 1 shows the influence of lightweight compression on query processing time including input data reading, processing and creating output. The bar on the left (CPU) presents the basic query processing pipeline with three activities: reading, processing and creating output. (GPU) bar shows processing time on GPU processing which additionally requires some time for data copying. Processing time is much shorter but additional copying makes overall speed-up not so impressive. The next column presents the same configuration but with lightweight compression of the data. Now copying time is much shorter. Thanks to GPU abilities data decompression time is not influencing the overall time noticeably. In the contrary the same approach but run purely on CPU suffers from long decompression time (the last column).

Our previous work on time series compression problems [19–21] and GPU utilization in time series processing showed that GPU may be successfully introduced

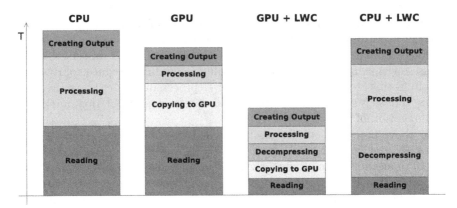

Fig. 1. General query processing time influenced by lightweight compression and GPU processing.

as a database coprocessor if accompanied by a lightweight compression of the stored data. We created a dynamic compression planner which was able to combine several lightweight compression methods in order to achieve the best compression results. However, the optimal compression may not always be acceptable due to possibly long decompression time.

In this work we extend the previous findings by a new multi-objective compression planner which may find a near-optimal[1] multi objective compression plan.

Section 2 presents a general view of the system including time series data model and data flow. Section 3 describes used lightweight compression methods which are suitable for fast GPU processing. Sections 4, 5 and 6 contain the main contribution of our work: the dynamically optimized compression system including plans estimation and bi-objective plan selection. Experimental runtime results are contained in Sect. 7 while Sect. 8 contains final remarks and conclusions.

1.1   Motivation and Related Work

Compression of time series is an interesting and widely analysed computational problem. Lossless methods often use some general purpose compression algorithms with several modifications according to knowledge gathered from data. On the other hand, lossy compression approximate data using, for instance, splines, piecewise linear approximation or extrema extraction [14]. For industrial monitoring systems, lossy compression cannot be used due to possible degradation of anomalies.

An important challenge is to improve compression factor with an acceptable processing time in case of variable sampling periods. Interesting results in the

---

[1] In this work we understand optimal compression as the best compression within available lightweight algorithms.

filed of lossless compression done on GPU were presented by Fang et al. [13]. Using a compression planner it was possible to achieve significant improvement in overall query processing on GPU by reducing data transfer time from RAM to global device's memory space. The strategy applied in our work is based on statistics calculated from inserted data and used to find an optimal cascaded compression plan for the selected lightweight methods.

The GPU compression topic was raised in several studies. Interesting results on GPU compression where presented by Andrzejewski et al. [5] where Word Aligned Hybrid compression algorithm for GPU was presented. Wu et al. [24] discussed implementation of Lempel-Ziv 77 (LZ77) algorithm on CUDA framework and showed that time complexity of this algorithm was to high on GPU processor when compared to CPU classical implementation. This was caused by too many branches in the algorithm which are not suited well for CUDA model of parallelism.

In a time series database we often observe data grouped into portions of very different characteristics. Optimal compression should be able to apply different compression plans for different time series and different time periods.

In case of lossless compression one can use common algorithms (ZIP, LZO) which tend to consume lot of computation resources [6,26] or lightweight methods which are faster but not so effective. Dynamic composition of several compression methods may improve this significantly by combination of properties of both approaches: it is lossless but much faster than common algorithms, offers acceptable compression ratios and may be computed incrementally. Selection of an optimal strategy (among available lightweight compression algorithms) is done upon data statistical information.

However, the challenge of multiprocessor and multi-GPU computational nodes raise another question: is it possible to improve these methods further including hardware specific information and estimated decompression time? In this work we extend our previous findings by a new multi-objective compression planner which may find an optimal compression plan under compression ratio and decompression speed optimization objectives. A database system will be able to benefit by better estimation of time constraints for query execution.

2   Time Series Database System with Compressed Storage

A typical time series database consists of three layers: data insertion module, data storage and querying engine. This section presents a general view of a prototype heterogeneous time series database system developed as a test-bed for our compression algorithms and optimization methods. It uses GPU as a coprocessor for database operations and data compression. Optimal resources (GPU and CPU) utilization requires a heterogeneous query planner which is addressed in another paper [22]. Here, we shall only focus on assuring optimal compression for this system.

2.1   Time Series Data Model

The data acquisition from ongoing measurements, industrial processes monitoring [15], scientific experiments [23], stock quotes or any other financial and
business intelligence sources has got continuous characteristic. These discrete
observations T are represented by pairs of a *timestamp* and a *numerical value*
(t_i, v_i) with the following assumptions:

- number of data points (timestamps and their values) in one time series should
  not be limited;
- each time series should be identified by a name which is often called a *metric
  name*;
- each time series can be additionally marked with a set of *tags* describing
  measurement details which together with metric name uniquely identifies time
  series;
- observations may not be done in constant time intervals or some points may
  be missing, which is probable in case of many real life data (Fig. 2).

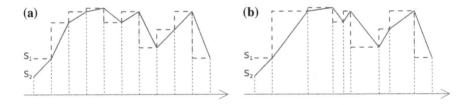

Fig. 2. Time series. (a) fixed time measurements (b) variable time measurements. Characteristics of the plot (s_1 – piecewise constant, s_2 – piecewise linear) depends on the
interpretation of the measured data value.

The last assumption is important since industrial applications often cannot
guarantee either constant measurement period or correct measurement and data
transfer.

Our prototype system does not limit possible data which can be inserted
and analysed. The only requirement of our system is that data must have a
form of *time series*, which we understand as a collection of observations made
sequentially in time [9].

In the presented data model known from for example OpenTSDB [2] one time
series is uniquely identified with a metric name and a set of tags. Combination of
tags let a user express many different queries in a very simple way. For instance
for the input data (timestamp, metric name, value, tags):

```
1386806400 cpu.load 0.20 node=alpha type=system
1386806400 cpu.load 0.10 node=alpha type=user
1386806401 cpu.load 0.30 node=alpha type=system
```

```
1386806401 cpu.load 0.20 node=alpha type=user
1386806400 cpu.load 0.05 node=beta type=system
1386806400 cpu.load 0.10 node=beta type=user
1386806401 cpu.load 0.05 node=beta type=system
1386806401 cpu.load 0.40 node=beta type=user
```

we could issue a query for an overall average *system* type processes processor load for all known nodes for one day by:

```
q?start=2013-12-12:00:00&
   end=2013-12-12:23:59&m=avg:cpu.load{type=system}
```

receiving:

```
1386806400 cpu.load 0.135 node=alpha type=system
1386806401 cpu.load 0.165 node=alpha type=system
```

or for a maximum processor load among all known nodes and processes types:

```
q?start=2013-12-12:00:00&
   end=2013-12-12:23:59&m=max:cpu.load
```

receiving:

```
1386806400 cpu.load 0.20
1386806401 cpu.load 0.40
```

2.2 Data Insertion

In this section we briefly describe data flow in our system, which is composed of three layers: data insertion, long term storage and data retrieval.

The insertion layer is preceded by a set of collectors which gather data from sensors, probes or other sources. These collectors sending data to the data insertion interface are considered external and beyond the scope of this paper.

The general view of our system's basic components and data flow is not different from other databases. The noticeable extension includes GPU and CPU processing. The main data flow indicates: data collection, data buffering, data storing and data querying. Data buffering may use compression on GPU or CPU side. Obviously, CPU compression usually does not require any additional data transfers since Data Buffer and CPU compression may be done by the same device and within the same memory space. However, compression performed by GPU requires extra time for data transferring between RAM and GPU device's global memory. After the data is compressed it may be sent to the database daemon which inserts them into a long term storage.

Similar situation occurs during data retrieval and query evaluation. Fragments of data required for particular query need to be decompressed, filtered and transformed according to the query parameters. All these operations may be done within CPU or with GPU used as an external coprocessor.

Figure 3 indicated the possible transitions and data flow between components in our database system. Each transition may have non zero data transfer time. Each node may transform data changing their size but also consuming time. A sample data insertion procedure could involve the following path: Data Collector, Data Buffer, GPU Compression, Data Buffer, Database Daemon, Storage. Data retrieval contains more nodes, transitions and possible paths. For example, a path of a query evaluated on CPU but with data decompressed on GPU will contain the sequence: Storage → Query Engine → GPU Decompression → Query Engine → CPU Query Processing → Query Engine → Client. Selection of an optimal query plan must involve distributed data processing on heterogeneous devices which we addressed in [22].

Minimisation of data transfer time in a heterogeneous database system is the main driver for the research presented in this paper. Our approach focuses on finding the best possible compression method suited for certain incoming data but from two points of view: compression ratio and decompression time. Both these goals are crucial for current time series database systems.

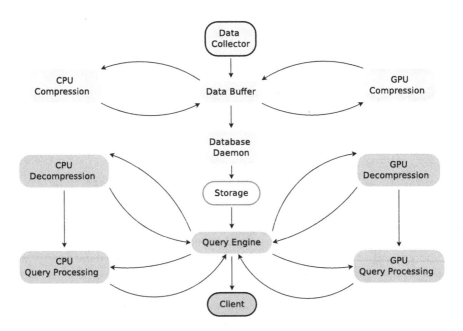

Fig. 3. A simplified data flow diagram including data insertion (light grey), data retrieval (dark grey) and storage (white) layers. Each transition may introduce additional data transfer cost. Each node may transform data influencing their size.

Time Series Storage. One of the most important properties of a time series database system is high performance and scalability. In many industrial solutions these assumptions lead to an architecture based on *Big Table* [8] and Map Reduce [11] applications.In such case time series data are stored at two different levels in the database:

– Data buffer – acts as a buffer for new data which enter the database. Data from the data buffer are periodically compacted, compressed and sent to the data archive. The buffer is composed of two separated data blocks: first one with timestamps and second one with values. This allows us to use different compression plans for both blocks.
– Permanent Storage – works as a long term data archive based on key-value table architecture. Metric name, tags and starting time are encoded in row key and column key.

Flow of Input Data. The data acquisition from ongoing measurements, industrial processes monitoring [15], scientific experiments [23], stock quotes or any other financial and business intelligence sources has got continuous characteristics. We assume that data collectors keep sending data to the system all the time and the system must respond in the real-time. Tight efficiency constraints must be met in order to assure that the data will not wait before being consumed for unacceptably long time.

Due to optimization purposes, data sent to the data storage should be ordered and buffered into portions, minimizing necessary disk operations but also minimizing the distributed storage nodes intercommunication. Buffering also prepares data to be compressed and stored optimally in an archive. Simplicity of data model imposed separated column families for compressed and raw data. Time series are separately compacted into larger records (by a metric name and tags) containing a specified period of time (e.g. 15 min, 2 h, 24 h – depending on the number of observations). This step directly preceded dynamic compression described in the next sections.

Finally, when a single record in a buffer is compressed and ready to be send to the long term permanent storage it is flushed and delivered to a NoSQL database which processes it according to its internal rules.

2.3 Data Retrieval

The last important part of the system is the query engine responsible for user-database interactions.

Execution of database queries is an example of a successful application of GPU co-processors which may accelerate numerous database computations, e.g. relational query processing, query optimization, database compression or supporting time series databases [7,20,21].

Distribution of workload between numerous CPU and GPU devices require careful planning of query execution strategy including not only data transfer costs but also device load, its efficiency or even energy consumption. In our previous publication we elaborated on bi-objective query planner which achieved interesting results in case of a heterogeneous query planning [22].

For the purposes of this work we indicate that the influence of compression methods used in a data storage on query evaluation is twofold. First, it may dramatically reduce data transfer time between system components and second, it may increase query evaluation time by additional decompression.

2.4 Searching for Optimal Compression

In any database system the data transfer time between distributed nodes or components significantly influences the overall performance of the system. This situation may be partially improved by compression but only if its additional cost is justified by gained speed up. Fine tuned lightweight compression methods offer interesting compression ratio with acceptable performance, especially if used on a GPU device [20,21].

In order to select an optimal compression method one must consider the following factors:

- Predicted compressed buffer size
- Predicted compression and decompression time
- Computational resources needed
- Additional method's properties.

Achieving best possible compression ratio and shortest possible working time are two contradicting objectives. Thus, a definition of an optimum solution set should be established. In this paper we use the predominant Pareto optimality [16].Given a set of choices and a way of valuing them, the Pareto set consists of choices that are Pareto efficient. A set of choices is said to be Pareto efficient if we cannot find a reallocation of those choices such that the value of a single choice is improved without worsening values of others choices.As bi-objective optimization is NP-hard, we need an approximate solution [18].

Our compression planner computes the best compression scheme upon all available algorithms, knowing their properties and input data characteristics.

3 Lightweight Compression Algorithms

In this section we present the compression algorithms and their modifications for the parallel execution on a GPU. Detailed description of presented compression algorithms may be found in [12,13,20,26].

Patched Lightweight Compression. The main drawback of many lightweight compression schemes is that they are prone to outliers in the data frame. For example, consider following data frame $\{1, 2, 3, 2, 2, 3, 1, 1, 64, 2, 3, 1, 1\}$, one could use the 2 bits fixed-length compression to encode the frame, but due to the outlier (value 64) we have to use 6-bit fixed-length compression or more computationally intensive 4-bit dictionary compression. Solution to the problem of outliers has been proposed in [26] as a modification to three lightweight compression algorithms. The main idea was to store outliers as exceptions. Compressed block consists of two sections: the first keeps the compressed data and the second exceptions. Unused space for exceptions in the first section is used to hold the offset of the following exceptions in the data in order to create linked list, when there is no space to store the offset of the next exception, a *compulsive exception* is created [26]. For large blocks of data, the linked lists approach may fail because the exceptions may appear sparse thus generate a large number of compulsory exceptions. To minimise the problem various solutions have been proposed, such as reducing the frame size [26] or algorithms that do not generate compulsive exceptions [12, 25]. The algorithms in this paper are based largely on those described by Yan [25]. In this version of the compression block is extended by two additional arrays - exceptions position and exceptions remainders values (i.e. the remaining bits). Decompression involves extracting data using the underlying decompression algorithm and then applying a patch (from exceptions remainders array) in the places specified by the exceptions positions. As exceptions are separated, data patching can be done in parallel. During compression, each thread manages two arrays for storing exception values and positions. After compression, each thread stores exceptions in the shared memory, similarly exceptions from shared memory are copied to the global memory. Patched version of algorithms are only selected if compression ratio improves. Otherwise non patched algorithms are used. Therefore complex exceptions treatment may be omitted speeding up the final compression.

Float to integer scaling (SCALE). Converts float values to integer values by scaling. This solution can be used in case where values are stored with given precision. For example, CPU temperature 56.99 can be written as 5699. The scaling factor is stored in compression header.

Differential representation (DELTA). Stores the differences between successive data points in frame while the first value is stored in the compression header. Works well in case of sorted data, such as measurement times. For example, let us assume that every 5 min the CPU temperature is measured starting from 1367503614 to 1367506614 (Unix epoch timestamp notation), then this time range may be written as $\{300, \ldots, 300\}$.

(Patched) Fixed-length Minimum Bit Encoding (PFL and FL). FL and PFL compression works by encoding each element in the input with the same number of bits thus deleting leading zeros at the most significant bits in the bit representation. The number of bits required for the encoding is stored in the compression header. The main advantage of the FL algorithm (and its variants) is the fact that compression and decompression are highly effective on GPU because these routines contain no branching-conditions, which decrease parallelism of SIMD operations. For the best efficiency dedicated compression and decompression routines are prepared for every bit encoding length with unrolled loops and using only shift and mask operations.Our implementation does not limit minimum encoding length to size of byte (as in [13]). Instead each thread (de)compresses block of eight values, thus allowing encoding with smaller number of bits. For example, consider following data frame $\{1, 2, 3, 2, 2, 3, 1, 2, 3, 1, 1\}$, one could use the 2 bits fixed-length compression to encode the frame.

(Patched) Frame-Of-Reference (PFOR and FOR). Works similarly to FL and PFL, except before compression it transforms each value into an offset from the reference value (for example smallest value) in compression block. Reference value is then stored in compression header. In this situation, we need exactly $\lceil \log_2(\max - \min + 1) \rceil$ bits to encode each value in the frame.

For example, this is useful when storing measurement times, consider time range $\{1367503614, \ldots, 1367506614\}$, then using for we only need $\lceil \log_2(1367506614 - 1367503614 + 1) = 12 \rceil$ bits to store each value in this range (as opposed to 31 bits without this transformation).

(Patched) Dictionary (DICT and PDICT). DICT is suitable for data that have only a small number of distinct values. It uses a dictionary of distinct values. For compression and decompression purposes, dictionary is loaded into the shared memory. Binary search is used during compression to lookup values, then an index of value is used to encode. Decompression simply retrieves values at given index from dictionary. DICT writes indexes using byte-aligned types, for better compression a combination with other compression algorithm should be used. For example, consider data frame $\{0, 500, 1500, 100, 100, 1500000, 100, 15000\}$ using DICT only 1 byte is needed to store each value (even less if combined with other compression algorithm) in comparison to pure FL where more than 2 bytes would have been used.

Run-Length-Encoding (RLE) and Patched Constant (PCONST). RLE encodes values with a pair: value and run length, thus using two arrays to compress data. Consider following data frame $\{1, 1, 1, 1, 1, 2, 2, 2, 2, 3, 3, 3\}$, then RLE would create two arrays: values $\{1, 2, 3\}$ and run length $\{5, 4, 3\}$. PCONST is a specialized version of RLE where almost whole data frame consist of one value with some exceptions. This may be reconstructed using: frame length, constant value and PATCH arrays. For example, let us assume that a measurement is done every five minutes with some exceptions, then delta is almost always constant and equals 300, any other value will be stored as exception.

4   Cascaded Compression Planner

The goal of this part of the system is to find suitable cascaded compression plans for the data gathered in the input buffer. It is composed of three parts:

- selection of suitable cascaded compression plans – mainly based on the specifics of the algorithms and the characteristics of the data set (see rest of this Section).
- evaluation of selected cascaded compression plans – based on the dynamic statistics generator and compressed data size estimation (see Sect. 5).
- bi-objective plan selector – which uses decompression time estimation and compressed data size to choose the final one (see Sect. 6).

Cascaded compression can significantly improve the compression ratio. However, searching for the most efficient compression method even for relatively short plans composed for several compression steps (i.e. using 6 compositions out of 10 algorithms with repetitions) may generate a very large search space (in our example $\sum_{i=1}^{6} 10^i = 1,111,110$). Due to tight time constraints we proposed a reduction of this problem by static planner and hints system.

Note that in fact, the situation is even more complicated, because of possible compression algorithms parametrization, e.g. (P)FL and (P)FOR take as an argument number of j bits used to encode each value, where $1 \leq j \leq 32$. This topic will be discussed with details in Sect. 5.

4.1 Reduction of Compression Plans Search Space: Static Planner

In the first static stage we determined acceptable transitions between compression algorithms which were divided into three categories: $\mathcal{T}$ – initial transformation, $\mathcal{B}$ – base compression, $\mathcal{H}$ – helper compression. The complete compression schema is always composed of algorithms selected from these subsequent categories $\bar{P} \subseteq P = \{(t, b, h) : t \in \mathcal{T}, b \in \mathcal{B}, h \in \mathcal{H}\}$, with the following purposes:

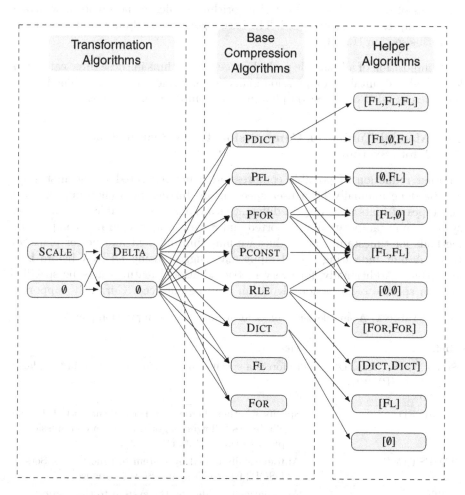

Fig. 4. The composition graph of all available compression plans within the given assumptions. Helper auxiliary algorithms are applied to additional arrays (from one to three) returned by the base compression algorithms.

1. $\mathcal{T}$ – **Transformation algorithms (SCALE, DELTA).** Improve properties of data storage and prepare for better compression. All algorithms in this section are optional but may be used together (if present must be applied in the given order).
2. $\mathcal{B}$ – **Base compression algorithms (PDICT, PFL, FOR, FL, DICT, PFOR, RLE, PCONST).** Only one algorithm may be selected as the base algorithm. All algorithms in this section generate from one up to three resulting arrays. Some of the resulting arrays, may qualify for further compression using *Helper compression algorithms.*
3. $\mathcal{H}$ – **Helper compression algorithms (FOR, FL, DICT).** The algorithms used to compress selected arrays from the previous step. Each of the resulting arrays can be compressed with only one algorithm. In order to minimize the stages of decompression PATCH algorithms, which could create new arrays for compression, are excluded. The base algorithm used may limit algorithms in this section (Fig. 4).

Composition of all sensible paths between algorithms in these three categories leaves only 76 suitable compression plans out of former one million. The longest possible cascaded compression plan may be composed of six steps.

4.2 Manual Tuning of Compression Plans Search Space: Hints System

Another reduction of possible compression plans generated in the first stage can be done manually by a user speeding up further plan choosing. Number and types of hints may vary in different situations. For example, in time series systems timestamps are always sorted and if we consider separated compression methods for timestamps and values we may find different and better plans for them. A hint indicating sorted input may suggest using DELTA before base algorithms. Additionally, for every metric additional features may be specified or even specific compression algorithm may be enforced. Currently supported

Table 1. A sample set of hints for a time series compression planner.

Hints	Meaning
SCALE, (P)FL, RLE, DELTA, (P)FOR, (P)DICT, PCONST	Enforces a specific compression algorithm in the plan
DSORTED	Specify whether the data is distinct and sorted. If true eliminates following algorithms from compression plan: PCONST, RLE, (P)DICT
TIMESTAMP	Automatically added by system to timestamps. Sets DSORTED to True and SCALE to False
DATA	Automatically added by the system to time series values. If not specified otherwise it sets DSORTED to False and SCALE to False

hints are located in Table 1. After this step we select a subset $P' = P'(D) \subseteq \bar{P}$ of compression plans, where D – is data set to be compressed, and $P'(D)$ is a reduced subset of $\bar{P}$ after using hints elimination.

5   Dynamic Statistics Generator

5.1   Finding the Optimal Parametrization and Compression Size Estimation

In this step, a plan with possibly maximal compression ratio is selected. In order to perform this task the system uses statistics and estimations computed dynamically from incoming data stream, for each metric and rolled time period separately. Pre-computing them and storing aside is not an optimal solution due to necessity of constant update and allocation of additional memory. Please note that if a plan contains an initial transformation algorithm it must be applied before calculating statistics because it influences data (Fig. 5).

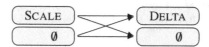

Fig. 5. Possible transformation algorithms and their composition.

Estimation results heavily depend on compression algorithms parameters. In [13] the choice of optimal parameters was straightforward, because the used algorithms supported only compression of value to byte-aligned size (which reduced number of parameters) and did not allow exceptions in data (only one set of parameters was correct). However, in compression algorithms and compression plans which use patching mechanism, optimal parameter selection is more complex. Factors such as the number of generated exceptions and estimated exception compression size should be taken into account. For example, the following data frame $\{1, 2, 3, 2, 32, 3, 3, 1, 64, 2, 1, 1\}$ could be compressed using PFL algorithm using 2 bits, 5 bits or 6 bits fixed-length, generating two exceptions $(32, 64)$, one exception (64) or no exceptions respectively. In this case, for each compression plan (selected in previous stages) a satisfactory set of parameters should be selected in order to correctly estimate compressed data size.

Recall that $P'(D)$ is a reduced subset of P after using static planner and hints elimination. Let $P'(D) \ni p = (t, b, h)$ be a cascaded compression plan, where $t \in \mathcal{T}, b \in \mathcal{B}, h \in \mathcal{H}$ are transformation, base and helper algorithms, respectively. A compression plan may be also written in simplified notation, i.e. ((SCALE, DELTA), PFL,(FL,FL)).

Let us denote the data after applying the transformation algorithms by $t(D)$ where $t \in \mathcal{T}$ transforms data D. Now, let $J(p, D)$ estimate compressed size of data D after applying compression plan p, i.e.

$$J(p, D) := J_b(h, t(D)) \qquad (1)$$

where $\bar{P} \ni p = (t, b, h)$ is a compression plan from subset of compression plans suitable for data D and J_b is estimation function for base algorithm b (see rest of this Sections for details). A pseudo code of an optimal compression plan selection is presented in listing SelectOptimalCompressionPlan. The rest of this section presents functions J_b for $b \in \{(P)\text{FL}, \text{RLE}, \text{DELTA}, (P)\text{FOR}, (P)\text{DICT}, \text{PCONST}\}$. This description is rather technical and not interested reader may skip to Sect. 6.

Procedure. SelectOptimalCompressionPlan(D)

    Input: D
    Result: $Plan$

```
1 P' = P'(D) ;  /* select reduced subset of P after using static planner
  and hints elimination */
2 min_size = size of data D;
3 p* = ∅;
4 for (t,b,h) = p in P'(D) do
5 |   D' = t(D);
6 |   if J_b(h,D') > min_size then
7 |   |   min_size = J_b(h,D');
8 |   |   p* = p;
9 |   end
10 end
11 return p*, p* optimal parameters;
```

Table 2. Symbols used in the definition of parametrization optimization

Symbol	Description
D	Dataset
$\#D$	Dataset length
$min(D)$	$\min_{d \in D} d$
$t_{sub}(m, D)$	Subtract m from each $d \in D$
B_{dict}	Number of bits of type used to encode dictionary keys
B_{base}	Number of bits of base type (i.e. 8 bits, 16 bits, 32 bits, 64 bits)
$b_{index}(D)$	The minimum number of bits required to store any number between 0 and $\#D$
$b_{min}D)$	The minimum number of bits required to store any value $d \in D$
$c_{in}(j, D)$	Compressed data size (in bits) of patch index when using j bits FL coding to compress data D
$c_{re}(j, D)$	Compressed data size (in bits) of remainders values when using j bits FL coding to compress data D

5.2 Notation

Most of the notation is gathered in Table 2. For example, B_{dict} represents number of bits of type that is used to encode dictionary keys and B_{base} number of bits of base type (i.e. 8 bits, 16 bits, 32 bits, 64 bits). Please note that in order to simplify the following formulas notation we deliberately omitted the fact that the resulting data should be aligned to a byte.

5.3 Optimal Parametrization and Compression Size Estimation for (P)FL and (P)FOR

For algorithms (P)FL and (P)FOR it is crucial to determine the optimal number of bits needed to compress data D. To estimate size of data compressed with (P)FL and (P)FOR algorithms we use *Bit histogram* statistic. Let us define *Bit histogram* as $s_{bit}(j, D) = \#\{d \in D : j$ bits are sufficient to write $d\}$ for $1 \leq j \leq$ 32 1. It is implemented on GPU using double buffering (registers and shared memory) parallel histogram scheme (Fig. 6).

Now, let b_{min} be the minimum number of bits required to store any value $d \in D$ and let $b_{index}(D)$ be the minimum number of bits required to store any number between 0 and $\#D$, i.e.

$$b_{min}(D) := \max_{1 \leq j \leq 32} \{j : s_{bit}(j, D) \neq 0\}, \tag{2}$$

$$b_{index}(D) := \lceil \log_2 \#D \rceil. \tag{3}$$

Consider a compression plan $p = (t, \text{FL}, h) \in P'$, i.e. a plan, where base compression algorithm b is set to FL. Since FL algorithm uses exactly $b_{min}(D)$ bits to compress each value in D, thus estimated compression size equals to

$$J_{FL}(h, D) := b_{min}(D) \cdot \#D, \tag{4}$$

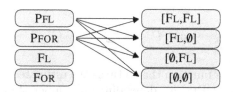

Fig. 6. PFOR, FOR, PFL and FL base algorithms and a possible composition with auxiliary algorithms.

(in case of compression plans with $b = \text{FL}$, helper algorithms are set to $h = \varnothing$). Clearly,

$$j_{FL} := b_{min}(D) \tag{5}$$

is the optimal parameter for FL.

Let now $p = (t, \text{FOR}, h) \in P'$. The estimation differs from the previous case, as now a transformation must be applied to data D. Let us define

$$t_{sub}(m, D) := \{d_i - m \colon d_i \in D\}, \tag{6}$$

which subtracts the value m from all values $d \in D$. To estimate the size of data after applying a compression plan $p = (t, \text{FOR}, h)$, we use

$$J_{FOR}(h, D) := J_{FL}(h, t_{sub}(min(D), D)), \tag{7}$$

i.e. the reference value here is $min(D)$ (in case of compression plans with $b = \text{FOR}$, helper algorithms are again set to $h = \varnothing$). Similarly to the previous case

$$b_{min}(t_{sub}(min(D), D)) \tag{8}$$

is the optimal parameter for FOR.

Before we consider the cases of patching algorithms in compression plans, we need to define helper functions c_{in} and c_{re}:

$$c_{in}(h_{in}, D) := \begin{cases} b_{index}(D) & \text{if } h_{in} = FL, \\ B_{base} & \text{otherwise,} \end{cases} \tag{9}$$

$$c_{re}(j, h_{re}, D) := \begin{cases} (b_{min}(D) - j) & \text{if } h_{re} = FL, \\ B_{base} & \text{otherwise.} \end{cases} \tag{10}$$

The returned values depend on the value of $(h_{in}, h_{re}) = h \in \{\text{FL}, \emptyset\}^2$, which indicates whether a helper compression algorithm is used or not. If $h_{in} = \text{FL}$, then c_{in} returns the number of bits needed to store each element in position array and if $h_{re} = \text{FL}$, then c_{re} returns the number of bits needed to store remainders values (i.e. the original value$-j$ bits). Otherwise, in both cases B_{base} is returned, i.e. number of bits needed to represent base type.

Also, let $c_{out}(j, D)$ be the number of outliers generated when using j bits as base bit encoding, defined as:

$$c_{out}(j, D) = \sum_{l=j+1}^{32} s_{bit}(l, D). \tag{11}$$

Next, let us define a function that returns the estimated compression size for compression plan $p = (t, \text{PFL}, h)$, depending on number of bits j used to encode the values:

$$g_{PFL}(j, h, D) := \#D \cdot j + c_{out}(j, D) \cdot (c_{in}(h_i, D) + c_{re}(j, h_r, D)), \tag{12}$$

where $h = (h_{in}, h_{re})$. The number of used bits j, determines the number of outliers, as each value that needs more then j bits to be written, will be treated as an outlier. Having said that, the returned value depends on the base compression array size (i.e. $\#D \cdot j$), on the number of outliers generated (i.e. $c_{out}(j, D)$) and the way how they will be stored (i.e. $c_{in}(h_i, D) + c_{re}(j, h_r, D)$).

The function J_{PFL} given by

$$J_{PFL}(h, D) := \min_{1 \leq j \leq 32} g_{PFL}(j, h, D) \tag{13}$$

returns the estimated compression size for compression plan $p = (t, \text{PFL}, h)$. Notice that the smaller j is, the smaller the size of the base compression array becomes and the more outliers we have. This is why we need to minimize g_{PFL} over j. We call

$$j_{PFL} := argmin\ J_{PFL}(h, D) := \min_{1 \leq j \leq 32} \{j : g_{PFL}(j, h, D) = J_{PFL}(h, D)\} \tag{14}$$

the optimal parameter for J_{PFL}, helper algorithms h and data D.

Similarly, for a compression plan $p = (t, \text{PFOR}, h)$ we define

$$J_{PFOR}(h, D) := J_{PFL}(h, t_{sub}(min(D), D)), \tag{15}$$

which returns the estimated compression size. Then

$$j_{PFOR} := argmin\ J_{PFL}(h, t_{sub}(min(D), D)) \tag{16}$$

is the optimal parameter for J_{PFOR}, helper algorithms h and data D.

5.4   Optimal Parametrization and Compression Size Estimation for (P)DICT

PDICT works on dictionary counter array and uses it to build an optimal dictionary with exceptions (minimizing estimated compression size after applying PDICT algorithm and using FL helper algorithms). Therefore, PDICT generates three output arrays: base, indexes and remainders while DICT generates only two: base and indexes from which indexes may be further compressed with FL. For PDICT at least base and remainders arrays must be compressed using FL to improve compression ratio, compared to similar compression plan but with DICT as base algorithm. Indexes array compression is optional.

To estimate size of data compressed with (P)DICT algorithm we use *Dictionary counter* statistic. Let us define $s_{dict}(a, D) = \#\{i \in I : a = d_i \in D\}$ as a *Dictionary counter*. As a side effect this generates a dictionary for further usage if needed. Implemented on GPU with sort and reduction operations. Mostly constructed using thrust library.

Now, let d_{keys} be a set of unique values in D, i.e.

$$d_{keys}(D) := \{a : s_{dict}(a)(D) \neq 0, a \in D\} \tag{17}$$

and let $d_{top}(k, D) \subseteq d_{keys}(D)$ be such that $\#d_{top}(k, D) = k$ and for $a \in d_{top}(k, D), b \notin d_{top}(k, topD)$ we have $s_{dict}(a) \geq s_{dict}(b)$ (if there are more sets satisfying this condition, we pick one of them) (Fig. 7).

Let d_{head} returns size of dictionary header array (i.e. array that stores sorted $d_{keys}(D)$)

$$d_{head}(D) := B_{dict} \cdot \#d_{keys}(D). \tag{18}$$

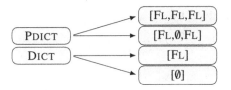

Fig. 7. PDICT and DICT base algorithms and possible composition with auxiliary algorithms.

Let us define helper functions d_{in}, d_{ba} and d_{re}:

$$d_{ba}(j, h_{ba}, D) := \begin{cases} j & \text{if } h_{ba} = FL, \\ B_{dict} & \text{otherwise,} \end{cases} \quad (19)$$

$$d_{in}(h_{in}, D) := c_{in}(h_{in}, D) \quad (20)$$

$$d_{re}(j, h_{re}, D) := \begin{cases} (b_{index}(\#d_{keys}(D)) - j) & \text{if } h_{re} = FL, \\ B_{dict} & \text{otherwise.} \end{cases} \quad (21)$$

The returned values depend on the value of $h \in \{FL, \emptyset\}^3$, which indicates whether a helper compression algorithm is used or not. If $h_{ba} = FL$, then d_{ba} returns the number of bits needed to store each element in base compression array position array, if $h_{in} = FL$ returns the number of bits needed to store each element in position (this is exactly the same as c_{in} in the previous section) and if $h_{re} = FL$, then d_{re} returns the number of bits needed to store remainders' values (i.e. the original value $- j$ bits).

Consider a compression plan $p = (t, \text{DICT}, h) \in P'$, i.e. a plan, where base compression algorithm b is set to DICT. Then estimated compression size equals to

$$J_{DICT}(h, D) := d_{ba}(h, D) \cdot \#D + d_{head}(D), \quad (22)$$

and includes size needed to store auxiliary dictionary table. Also in case of compression plans with $b = \text{DICT}$, helper algorithms may appear so this is also included.

The following formulas return estimated compression size for compression plan $p = (t, \text{PDICT}, h)$:

$$d_{out}(i, D) := \sum_{a \notin d_{top}(2^i, D)} s_{dict}(a, D) \quad (23)$$

$$G_{\text{PDICT}}(i, h, D) := d_{ba}(i, h_{ba}, D) \cdot \#D + d_{head}(D)$$
$$+ d_{out}(i, D) \cdot \left(d_{in}(h_{in}, D) + d_{re}(i, h_{re}, D) \right) \quad (24)$$

$$J_{\text{PDICT}}(h, D) := \min_{1 \leq i \leq log_2 B_{dict}} G_{\text{PDICT}}(i, h, D) \quad (25)$$

where $b = \text{PDICT}$, $h = (h_b, h_{in}, h_{re})$ and at least $h_{ba} = FL$, $h_{re} = FL$ (i.e. any compression plan in this form $(t, \text{PDICT}, (FL, h_{in}, FL))$). Without above assumption PDICT will not improve compression ratio compared to DICT. We call

$$jPDICT := argmin\ J_{\text{PDICT}}(h, D) := \min_{1 \leq i \leq log_2 B_{dict}} \{i\colon g_{DICT}(i, h, D) = J_{\text{PDICT}}(h, D)\}$$
(26)

the optimal parameter for J_{PDICT}, helper algorithms h and data D.

5.5   Optimal Parametrization and Compression Size Estimation for RLE and PCONST

To estimate size of data compressed with RLE algorithm we use *Run length counter*. Let us define $s_{rle}(D) = (a_r, a_v)$ where a_r and a_v are run-length and values arrays generated from data D, respectively. This is implemented on GPU with reduction operation on key-value pairs.

Let us define a function that returns the estimated compression size for compression plan $p = (t, RLE, h)$:

$$J_{\text{RLE}}(h, D) := B_{base} \cdot \#a_r + B_{base} \cdot \#a_v$$

where $(a_r, a_v) = s_{rle}(D)$. Now, additional helper algorithms may be used on a_r and a_v arrays, this however requires additional statistic generation step (Fig. 8).

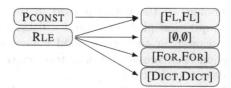

Fig. 8. RLE and PCONST base algorithms and possible composition with auxiliary algorithms.

Lastly, the PCONST algorithm, which in nature reminds RLE however, achieving the objective somewhat differently. In this algorithm a dominant value $d_{top}(1, D)$ is selected from dataset D (and stored in header), all other values are stored as outliers in PATCH arrays (index and remainders arrays). Following formula returns estimated compression size for compression plan $p = (t, \text{PCONST}, h)$:

$$J_{PCONST}(h, D) := B_{base} + \sum_{a \notin d_{top}(1, D)} s_{dict}(a, D) \cdot (c_{in}(h_{in}, D) + c_{re}(0, h_{re}, D)).$$
(27)

$$J_{PCONST}(h, D) := B_{base} + d_{out}(0, D) \cdot s_{dict}(a, D) \cdot (c_{in}(h_{in}, D) + c_{re}(0, h_{re}, D)).$$
(28)

Note that, we do not divide values when creating remainders array, instead we whole value (without dividing). This is because we only create PATCH arrays in this algorithm. Decompression involves creation of constant data set using dominant value and applying PATCH array afterwards.[2]

6   Bi-objective Compression Plan Selection

The lightweight compression algorithms, are primarily designed for applications favouring compression/decompression speed over compression ratio. Unfortunately decompression of cascaded compression plan is usually more computationally demanding then decoding a simple compression scheme. We are facing a dilemma, how much of the processing speed may be sacrificed to gain better compression ratio.

Since there is no clear right answer, we propose optimization model which is designed to support compression plan selection in the presence of trade-offs between decompression speed and compression ratio.

Section 5 describes how to estimate compressed data size according to a predicted compression plan. In Sect. 6.2 we will construct a function which estimates decompression speed of cascaded compression plan. Finally, in Sect. 6.3 we will discuss how to combine those two objectives.

6.1   Notation

Table 3 contains a summary of the notation used in this section. Assume a set of units U, a decompression plan set P and a dataset D. The goal is to choose such a compression plan that gives good compression ratio and yet allows fast decompression (i.e. compressed data transfer time + decompression time $\geq$ decompressed data transfer time).

As a result *Cascaded Compression Planner* returns set $P' \subset P$. Let us define:

$$f_t(p'_i, D) = T(u, p'_i, D), \tag{29}$$
$$f_r(p'_i, D) = J(p'_i, D) \tag{30}$$

where $p'_i \in P'$, and u is device on which time measurements where done.

6.2   Decompression Time Estimation

Compression ratio and decompression time are the input parameters to bi-objective compression optimization. Therefore, all candidate compression plans need to have their decompression time estimated. In this, section we will discuss the problem of estimation of decompression time for compression plans.

Estimation of decompression time for the candidate compression plan is calculated based on the estimated decompression time for each algorithm contained in in the plan. Recall that $P'(D) \ni p = (t, b, h)$ is a cascaded compression plan,

---

[2] Note that this is a certain simplification, i.e. instead $c_{re}(0, h_{re}, \bar{D})$ where $\bar{D}$ is dataset after removing all instances of dominant value.

Table 3. Symbols used in the definition of our optimisation model

Symbol	Description
$U = \{u_1, u_2, \ldots u_n\}$	Set of computational units available to process data
D	Dataset
$p_i \in P$	Decompression plan p_i from set of decompression plans P
$T(u, p, D)$	Estimated run time of the decompression plan p using device u on the data D
f_t	Estimated maximal run time for decompression plan
f_r	Estimated compression ratio
f_b	Estimated decompression run time and compression ratio bi-objective scalarization

where $t \in \mathcal{T}, b \in \mathcal{B}, h \in \mathcal{H}$ are transformation, base and helper algorithms, respectively. One can treat t, h as vectors of operations to perform, where operations are lightweight compression algorithms or empty operation $\varnothing$.

Let us denote by $T(p, D)$ the estimated decompression execution time for compression plan p and data D:

$$T(p, D) := \sum_{i=0,1} m_{t_i}(\#D) + T_b(h, D). \tag{31}$$

In the next paragraphs we describe the details of time estimation functions for different lightweight compression algorithms. Not interested readers may skip to the Sect. 6.3 not loosing the main contribution of our work.

Decompression Time Estimation Details. Slightly abusing notation, by $m_\varnothing(l) := 0$ and $m_\varnothing(j, l) := 0$ we denote execution time of empty operation used in transform and helper algorithms sections, respectively.

Next, let us define functions which return estimated decompression time for following algorithms SCALE, DELTA, FL, FOR and DICT. Depending on algorithm type estimation functions require different arguments. We have $m_{\text{SCALE}}(l)$, $m_{\text{DELTA}}(l)$, $m_{\text{FL}}(j, l)$, $m_{\text{FOR}}(j, l)$, $m_{\text{DICT}}(d, l)$, respectively, where l stands for data size, j is the bit length used in FL and FOR algorithms, d is the size of used dictionary.

Now, we define decompression estimation functions T_b for compression plan $p = (t, b, h)$ and data D. For $b \in \{\text{FL}, \text{FOR}, \text{DICT}\}$ we have:

$$T_{\text{FL}}(h, D) := m_{\text{FL}}(j_{\text{FL}}, \#D) \tag{32}$$

$$T_{\text{FOR}}(h, D) := m_{\text{FOR}}(j_{\text{FOR}}, \#D) \tag{33}$$

$$T_{\text{DICT}}(h, D) := m_{\text{DICT}}(\#d_{keys}(D), \#D) + m_{h_{ba}}(b_{min}(d_{keys}(D)), \#D), \tag{34}$$

where j_{FL}, j_{FOR} are number of bit used for base encoding (see Eqs. 5 and 8), d_{keys} and b_{min} are defined in Eqs. 17 and 2, respectively, h_{ba} is a helper algorithm for DICT.

For $b \in \{\text{PFL}, \text{PFOR}\}$ we need to define auxiliary functions first. Recall that for $b \in \{\text{PFL}, \text{PFOR}\}$, helper algorithms are of the following form $h = (h_{in}, h_{re})$ and functions c_{in}, c_{re} which return the number of necessary bits for helper algorithms are given by Eqs. 9 and 10. Let

$$m_{PFL}(j, o, h, D) := m_{h_{in}}(c_{in}(h_{in}, D), o) + m_{h_{re}}(c_{re}(j, h_{re}, D), o), \qquad (35)$$

$$m_{PFOR}(j, h, D) := m_{PFL}(j, h, D) \qquad (36)$$

where j is the number of bits used for base encoding, o is the number of outliers and D represents data set.

Similarly for $b = \text{PDICT}$, helper algorithms are of the following form $h = (h_{ba}, h_{in}, h_{re})$ and functions d_{ba}, d_{in}, d_{re} which return the number of necessary bits for helper algorithms are given by Eqs. 19, 20 and 21. Let

$$m_{PDICT}(j, o, h, D) := m_{h_{ba}}(d_{ba}(j, h_{ba}, D), o)$$
$$+ m_{h_{in}}(d_{in}(h_{in}, D), o) + m_{h_{re}}(d_{re}(j, h_{re}, D), o) \qquad (37)$$

where j is the number of bits used for base encoding, o is the number of outliers and D represents data set.

Then for any plan where $b \in \{PFL, PFOR, PDICT\}$ we define:

$$T_{\text{PFL}}(h, D) := m_{\text{FL}}(j_{\text{PFL}}, \#D) + m_{PFL}(j_{PFL}, c_{out}(j, D), h, D), \qquad (38)$$

$$T_{\text{PFOR}}(h, D) := m_{\text{FOR}}(j_{\text{PFOR}}, \#D) + m_{PFOR}(j_{PFOR}, c_{out}(j, D), h, D), \qquad (39)$$

$$T_{\text{PDICT}}(h, D) := m_{\text{DICT}}(2^{j_{\text{PDICT}}}, \#D) + m_{PDICT}(j_{PDICT}, d_{out}(j, D), h, D) \qquad (40)$$

where $j_{PFL}, j_{\text{PFOR}}, j_{\text{PDICT}}$ are optimal parameters for PFL, PFOR and PDICT (see Eqs. 14, 16 and 26), c_{out} and d_{out} return the number of a outliers (see Eqs. 11 and 23).

Similarly for PCONST, define

$$T_{\text{PCONST}}(h, D) := m_{CONST}(\#D) + m_{PFL}(0, d_{out}(0, D), h, D). \qquad (41)$$

Decompression time of RLE depends only on data length [13], let function $m_{\text{RLE}}(l)$ return estimated decompression time for RLE:

$$T_{\text{RLE}}(h, D) := m_{RLE}(D) + \text{decompression time of helpers algorithms used.} \qquad (42)$$

6.3   Bi-objective Compression Planner

We will use *a priori* articulation of preference approach which is often applied to multi-objective optimization problems. It may be realized as the scalarization of objectives, i.e., all objective functions are combined to form a single function. In this work we will use *weighted product* method, where weights express user preference [16]. Let us define:

$$f_b(u, p, D) = f_r(p, D)^{w_r} \cdot f_t(u, p, D)^{w_t}$$

where w_t and w_c are weights which reflect how important cost and time is (the bigger the weight the more important the feature). It is worth to mention that a special case with $w_t = w_r = 1$ (i.e., without any preferences) is equivalent to Nash arbitration method (or objective product method) [16].

We can also extend this to handle a set of devices U. This allows us to consider compression plans taking into account all devices on which the data may be decompressed. Let us define function f_m: $f_m(p, D) = (\max_{u_k \in U} t_t(u_k, p, D))^{w_t} + f_r(p, D)^{w_r}$ this function optimizes for the slowest device for each plan.

7   Preliminary Runtime Results

In this section we discuss effectiveness of the dynamic compression planner in the context of the resulting compression ratio. To fully evaluate the proposed compression framework, we still need to perform more experiments regarding the context of decompression speed for many other data sets and devices. The detailed processing time analysis including threads instruction throughput and effectively achieved memory bandwidth needs a lot of runtime trials and will be addressed in the next paper.

We compared effectiveness of a dynamic compression planner and a single static plan within the same CF (Column Family – portion of data rolled in a database) by running the prototype system on samples from a set of network servers monitoring. The data included memory usage, the number of exceptions reported, services occupancy time or CPU load. Data covered a sample of 20 days of constant monitoring and contained about 91 K data points in a few time series being a sample from a telecommunication monitoring system.

We used the following equipment: *Nvidia® Tesla C2070 (CC 2.0)* with 2687 MB; 2 x Six-Core processor *AMD® Opteron^{TM}* with 31 GB RAM, *Intel®* RAID Controller *RS2BL040* set in RAID 5, 4 drives *Seagate® Constellation ES ST2000NM0011* 2000 GB, Linux kernel 2.6.38–11 with the CUDA driver version 5.0.

7.1   Evaluation of the Compression Planner

The evaluation was divided into two parts. The first measured efficiency of the dynamic planner and was intended to prove the basic contribution of this work. The second checked efficiency of GPU based statistics evaluation when compared to CPU and proved contribution concerning time efficiency.

7.2   Dynamic Compression Planner Evaluation

Figure 9 shows compression ratio (original size/compressed size) using several static plans (one compression plan for the whole column family) and a dynamic plan (dynamically chosen compression plan for different metrics, tags and time ranges). In case of timestamps, five static plans were generated using DELTA algorithm combined with five base compression methods (and helper compression algorithms if suitable). Similarly, for data values five plans where selected except

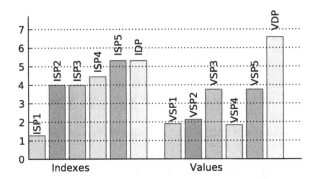

Fig. 9. Efficiency of the prototype dynamic compression system working on GPU for sample time series. Compression ratio (original size/compressed size) for static (*SP) and dynamic (*DP) plans. I stands for indices and V for values.

SCALE was used instead of DELTA. We may observe, that for timestamp arrays, compression ratio of a dynamic compression plan was equivalent to best static compression plan. This situation appeared because all time series were evenly sampled in this case. Therefore one static plan for all metrics generated the same results as a dynamic plan, selected for each time series separately. Note that in real systems, some measurements may be event-driven and thus dynamic plan could generate better results.

For data values, a dynamic compression plan almost doubles compression ratio of the best static compression plan which means that dynamic tuning was much better than selection of one static plan for the whole buffered column family. Obviously, this is heavily data dependant, but as a general rule a dynamic compression plan will never generate a compression plan worse than the best static plan (as it always minimizes locally). Additionally hints system may be used to enforce a static compression plan for cases when a dynamically generated compression plan does not produce satisfactory profits.

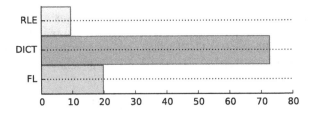

Fig. 10. Statistics calculation speed-up on sample data with 8 millions values compared to single threaded GPU. This includes GPU memory transfer (higher is better).

7.3   Evaluation of Statistics Calculations and Bandwidth of Compression Methods on GPU

In Fig. 10 on the right GPU statistic generator is compared to similar CPU version (implemented as a single thread). Generated statistics are then used to support the compression planner selection. A significant speed-up of factors from 10 to 70 was gained thus usage of GPU platform in statistics calculation step is justified.

Furthermore, GPU platform allows to archive high compression bandwidth for lightweight compression schemes (see Table 4 and results from [13]). We conclude that GPU may be used just as a kind of compression coprocessor even if there are no other computations done on a GPU side.

Table 4. Achieved bandwidth of pure compression methods (no IO).

Algorithm	DELTA	SCALE	(P)DICT	(P)FOR	(P)FL	RLE	PCONST
GB/s	28.875	41.134	6.924	9.124	9.375	5.005	2.147

8   Conclusions and Future Research

Monitoring of complex computer infrastructure is already an important industrial problem. Time series database system try to address many of the problems which may appear, like: scalability, robustness, safety and availability. Our research focuses on a time series database system supported by GPU coprocessors which may be used for many purposes, from data compression to analysis and aggregation.

In this paper, touching lightweight compression methods we successfully extended results from [13,19]. We not only designed and implemented new patched compression algorithms on GPU (i.e. Patched DICT, Patched Const. and Patched Fixed Length) but also presented a dynamic compression planner. Our novel prototype system was adapted to time series compression in a NoSQL database. The compression method is composed of several nested algorithms.

Furthermore our compression planner uses dynamic data statistics calculated on the fly using a GPU device for the best possible lightweight cascaded compression plan selection. We believe that the resulting compression ratios and algorithms bandwidth (please refer to Table 4) combined with ultra fast decompression [13,19] on GPU are especially attractive for time series databases.

Our future work will concentrate on query optimization in hybrid CPU/GPU environment, query execution on partially compressed data and extending dynamic compression planner by introducing additional costs factors (i.e. decompression execution time [13] or potential of query execution on compressed data) leading to a full-flagged time series database system.

References

1. Apache HBase (2013). http://hbase.apache.org
2. OpenTSDB - A Distributed, Scalable Monitoring System (2013). http://opentsdb.net/
3. ParStream - website (2013). https://www.parstream.com
4. TempoDB - Hosted time series database service (2013). https://tempo-db.com/
5. Andrzejewski, W., Wrembel, R.: GPU-WAH: applying GPUs to compressing bitmap indexes with word aligned hybrid. In: Bringas, P.G., Hameurlain, A., Quirchmayr, G. (eds.) DEXA 2010, Part II. LNCS, vol. 6262, pp. 315–329. Springer, Heidelberg (2010)
6. Boncz, P.A., Zukowski, M., Nes, N.: Monetdb/x100: hyper-pipelining query execution. In: CIDR, pp. 225–237 (2005)
7. Breß, S., Schallehn, E., Geist, I.: Towards Optimization of Hybrid CPU/GPU Query Plans in Database Systems. In: New Trends in Databases and Information Systems, pp. 27–35. Springer, Heidelberg (2013)
8. Chang, F., Dean, J., Ghemawat, S., Hsieh, W.C., Wallach, D.A., Burrows, M., Chandra, T., Fikes, A., Gruber, R.E.: Bigtable: a distributed storage system for structured data. In: OSDI'06: Seventh Symposium on Operating System Design and Implementation, Seattle, WA, November, pp. 205–218 (2006)
9. Chatfield, C.: The Analysis of Time Series: An Introduction, 6th edn. CRC Press, Florida (2004)
10. Cloudkick. 4 months with cassandra, a love story, March 2010. https://www.cloudkick.com/blog/2010/mar/02/4_months_with_cassandra/
11. Dean, J., Ghemawat, S.: Mapreduce simplified data processing on large clusters. Commun. ACM **51**(1), 107–113 (2004)
12. Delbru, R., Campinas, S., Samp, K., Tummarello, G.: Adaptive frame of reference for compressing inverted lists. Technical report, DERI - Digital Enterprise Research Institute, December 2010
13. Fang, W., He, B., Luo, Q.: Database compression on graphics processors. Proc. VLDB Endowment **3**(1–2), 670–680 (2010)
14. Fink, E., Gandhi, H.S.: Compression of time series by extracting major extrema. J. Exp. Theor. Artif. Intell. **23**(2), 255–270 (2011)
15. Lees, M., Ellen, R., Steffens, M., Brodie, P., Mareels, I., Evans, R.: Information infrastructures for utilities management in the brewing industry. In: Herrero, P., Panetto, H., Meersman, R., Dillon, T. (eds.) OTM-WS 2012. LNCS, vol. 7567, pp. 73–77. Springer, Heidelberg (2012)
16. Marler, R.T., Arora, J.S.: Survey of multi-objective optimization methods for engineering. Struct. Mult. Optim. **26**(6), 369–395 (2004)
17. OpenTSDB. Whats opentsdb (2010–2012). http://opentsdb.net/
18. Papadimitriou, C.H., Yannakakis, M.: Multiobjective query optimization. In: Proceedings of the Twentieth ACM SIGMOD-SIGACT-SIGART Symposium on Principles of Database Systems, pp. 52–59. ACM (2001)
19. Przymus, P., Kaczmarski, K.: Improving efficiency of data intensive applications on GPU using lightweight compression. In: Herrero, P., Panetto, H., Meersman, R., Dillon, T. (eds.) OTM-WS 2012. LNCS, vol. 7567, pp. 3–12. Springer, Heidelberg (2012)
20. Przymus, P., Kaczmarski, K.: Dynamic compression strategy for time series database using GPU. In: New Trends in Databases and Information Systems. 17th East-European Conference on Advances in Databases and Information Systems, 1–4 September 2013 - Genoa, Italy (2013)

21. Przymus, P., Kaczmarski, K.: Time series queries processing with gpu support. In: New Trends in Databases and Information Systems. 17th East-European Conference on Advances in Databases and Information Systems, 1–4 September 2013 - Genoa, Italy (2013)

22. Przymus, P., Kaczmarski, K., Stencel, K.: A bi-objective optimization framework for heterogeneous CPU/GPU query plans. In: CS&P 2013 Concurrency, Specification and Programming. Proceedings of the 22nd International Workshop on Concurrency, Specification and Programming, 25–27 September 2013 - Warsaw, Poland (2013)

23. Przymus, P., Rykaczewski, K., Wiśniewski, R.: Application of wavelets and Kernel methods to detection and extraction of behaviours of freshwater mussels. In: Kim, T., Adeli, H., Slezak, D., Sandnes, F.E., Song, X., Chung, K., Arnett, K.P. (eds.) FGIT 2011. LNCS, vol. 7105, pp. 43–54. Springer, Heidelberg (2011)

24. Wu, L., Storus, M., Cross, D.: Cs315a: final project cuda wuda shuda: Cuda compression project (2009)

25. Yan, H., Ding, S., Suel, T.: Inverted index compression and query processing with optimized document ordering. In: Proceedings of the 18th International Conference on World Wide Web, pp. 401–410. ACM (2009)

26. Zukowski, M., Heman, S., Nes, N., Boncz, P.: Super-scalar RAM-CPU cache compression. In: ICDE'06. Proceedings of the 22nd International Conference on Data Engineering, pp. 59–59. IEEE (2006)

A Global Paradigm for Designing Parallel Relational Data Warehouses in Distributed Environments

Soumia Benkrid[1,2(✉)], Ladjel Bellatreche[1], and Alfredo Cuzzocrea[3]

[1] LIAS/ISAE-ENSMA, Poitiers, France
{soumia.benkrid,bellatreche}@ensma.fr
[2] National High School for Computer Science (ESI), Algiers, Algeria
[3] ICAR-CNR and University of Calabria, Rende, Italy
cuzzocrea@si.deis.unical.it

Abstract. Designing a *Parallel Relational Data Warehouse* (PRDW) consists of a set of tasks: (*i*) choosing the hardware architecture; (*ii*) fragmenting the data warehouse schema; (*iii*) allocating the generated fragments; (*iv*) replicating fragments in order to ensure high performance; (*v*) defining the strategies for load balancing and query processing. The major drawback of this life-cycle is the fact that it does not consider the inter-dependency among sub-problems related to the design of PRDW, and it makes use of heterogeneous metrics to evaluate the "quality" of the final design. In previous research efforts, we introduced an analytical cost model for parallel OLAP query processing in cluster environments. In a second experience, we have taken into account the inter-dependency existing between fragmentation and allocation. In this paper, we propose a novel methodology, called $\mathcal{F\&A\&R}$, which further extends previous results, and defines an approach where *the main PRDW design phases (i.e., fragmentation, allocation, and replication) are performed simultaneously, in a global fashion*. In particular, our approach determines whether the fragmentation pattern currently generated is relevant to the allocation process or not. An original method of supporting data replication, based on *fuzzy k-means clustering*, is also proposed and successfully integrated within the whole design framework. Finally, we experimentally assessed the performance of $\mathcal{F\&A\&R}$ against a well-known data warehouse benchmark, with very promising results.

Keywords: Data warehouse · Distributed environment · Fragmentation · Allocation · Replication · Load balancing · Analytical cost model · Design methodology

1 Introduction

Today volumes of data are increasing more and more due to the rise of new infrastructures, such as *Clouds* [1], and new devices, such as sensors [22]. On the other hand, social networks (e.g., Facebook, Twitter and LinkedIn) collect

© Springer-Verlag Berlin Heidelberg 2014
A. Hameurlain et al. (Eds.): TLDKS XV, LNCS 8920, pp. 64–101, 2014.
DOI: 10.1007/978-3-662-45761-0_3

billions of data bytes, and predicting the behavior of users in order to improve their services via analyzing so-collected large data volumes is becoming increasingly hard. As a consequence, traditional *Data Warehouses* (DW) have become obsolete and *Parallel Relational Data Warehouses* (PRDW), instead, have been proposed as a robust and scalable platform for storing, processing and analyzing large volumes of data within the layers of modern analytics infrastructures. Similarly, a large number of software companies are positioned around the market with the goal of providing Business Intelligence solutions on top of large volumes of data, such as Teradata[1], Netezza[2], and so forth. In line with these major trends, Small and Medium-sized Enterprises (SME) are defining new classes of jobs dealing with so-called *Big Data* such as Data Architect, Data Visualizer, Data Analyst etc., thus exposing a clear commercial demand. This despite Big Data software platforms still remain costly for SME in terms of license fees and costs of installation and maintenance (stirred-up by the current economic crisis).

Under a general view, designing a PRDW comprises the following main steps (see Fig. 1): (1) choosing the hardware architecture, (2) partitioning the target DW, (3) allocating the so-generated fragments over available nodes, (4) replicating fragments for efficiency purposes, (5) defining efficient query processing strategies, (6) defining efficient load balancing strategies. Currently, several types of hardware architecture are available, such as *Shared-Nothing*, *Shared-Disk*, *massively parallel processors* and *Clusters of workstations*. The Shared-Nothing architecture has been proposed by DeWitt [31] as the reference architecture for supporting high-performance data warehouses modeled in terms of relational star schemas. As the choice of the hardware architecture is influenced by price, high-performance features, extensibility and data availability [12], Clusters of workstations are very often used as a valid alternative to Shared-Nothing architectures (e.g., [5]).

According to this low-cost technology solution, the target DW is divided into disjoint units called *partitions* that do not introduce any loss or addition of information with respect to the corresponding combination of partitions kept in the original DW. Data partitioning can be done horizontally or vertically, alternatively. Horizontal partitioning is essentially used to design PRDW. Data allocation consists in placing generated fragments over nodes of a reference parallel machine. This allocation may be either *redundant* (with replication) or *non redundant* (without replication). Once fragments are placed, global queries are executed over the processing nodes according to parallel computing paradigms.

In more detail, parallel query processing on top of a PRDW (a critical task within the family of parallel computing tasks) includes the following phases: (*i*) rewriting the global query according to the fixed DW fragmentation schema; (*ii*) scheduling the evaluation of so-generated sub-queries over the parallel machine according to a suitable allocation schema. Generating and evaluating sub-queries such that the query workload is evenly balanced across all the processing nodes is the most difficult task in the parallel processing above.

---

[1] http://www.teradata.com/.

[2] www.ibm.com/software/fr/data/netezza/.

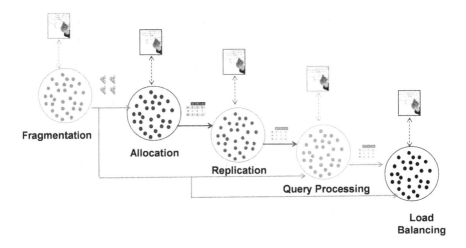

Fragmentation

Allocation

Replication

Query Processing

Load
Balancing

Fig. 1. Life-cycle of the PRDW design process

Indeed, load imbalance can be caused by one or possibly a combination of the following two phenomena: (*i*) *data skew*, which refers to the situation where data are unevenly distributed across the local memories of different processors – it usually occurs when the current data partitioning function makes use of attributes whose data value distributions are non-Uniform; (*ii*) *processing skew*, which is caused by the unpredictable nature of the processing itself and it may be propagated by the data skew at the beginning – it refers to the situation where a significant part of the workload is executed by a few processors while other processors are relatively idle.

Load balancing is usually performed by means of the so-called *multi-reordering* process. According to this process, multiple processors that have small average loads are selected in order to participate to the load balancing. Then, each free processor is moved as to becoming adjacent (according to the node network topology) to a high-loaded processor, the load of which is then shared with the (newly-introduced) free processor. This so-determined *data migration* task may cause high communication costs, which overall lower the global throughput of the PRDW architecture. From active literature (e.g., [2]), it is well-understood that communication cost is a factor that must be mastered depending on the available infrastructure, and that most of data access must be local (for efficiency purposes). Therefore, *data replication* has become a strict requirement of PRDW architectures in order to guarantee avoiding bottlenecks and reducing communication costs. To this end, replication aims at (*i*) ensuring data availability and fault tolerance, (*ii*) improving data locality by following the criterion of placing a job at the same node where its data are located, and (*iii*) achieving load balancing by distributing work across data replicas.

On the basis of the guidelines above, here we assert that PRDW design can be modeled as the following tuple: $\langle Arch, DP, DA, DR, LB \rangle$, where *Arch* denotes the parallel architecture, DP the data partitioning schema, DA the data allocation schema, DR the data replication schema and LB the load balancing scheme, respectively. Unfortunately, each one of the sub-tended problem of the main PRDW design problem is NP-hard [2,4,53].

1.1 Contributions of this Research

Under a broader vision, the PRDW design problem can be thought as a *set of services* offered by *actors* which communicate and cooperate among them in order to obtain a high throughput in the whole PRDW architecture. In our research, by exploiting this metaphor, we particularly introduce five actors: *Partitioner, Allocator, Replicator, LoadBalancer, ParallelQueryProcessor*, each one focusing on a particular PRDW design aspect. By inspecting the active literature, comprehensive surveys of state-of-the-art research on PRDW design issues exist, but still researchers focus the attention on PRDW issues in an isolated manner, without considering the inter-dependency among the different issues. In fact, some focus on the data partitioning problem (e.g., [48,55,58]), others on the data allocation problem (e.g., [2,4,47]), or the data replication problem (e.g., [18,39,44]), or the parallel query processing problem (e.g., [3,42,43]). As a consequence, two main limitations may penalize the PRDW design phase: (*i*) neglecting the inter-dependency among the different-but-related PRDW design issues, and (*ii*) adopting heterogeneous metrics in order to identify the "quality" of the final solution (indeed, each one of the process led by the five actors above is evaluated according to a different metrics to this end). As regards related efforts, few initiatives have investigated the inherent dependencies of the different aspects of the main PRDW design problem. These approaches integrate the fragmentation and allocation processes under the name of "data placement" [19,29]. Recently, industrialists and academics make use of parallel processing as a cost model to identify the quality of data placement schema [6]. On the other hand, Stöhr *et al.* are the pioneers who re-visited the PRDW design problem on a parallel shared-disk machine.

In order to fulfill the limitations of the main PRDW design problem highlighted above, in this paper we propose a novel method for designing PRDW over parallel machines, where the basic idea consists in *considering the interaction among the different aspects of the main PRDW design problem in order to use a unique cost model that smoothly unifies all phases, hence achieving a global approach* (see Fig. 2). The final goal consists in packaging the PRDW design issues as a unified process that cements PRDW design phases and increases the "omniscience" of the actors. Since data partitioning plays an important role in the whole PRDW design, we consider it as the first important step in this design. According to our design approach, during the fragmentation phase, the *Partitioner* should consider that the target RDW to be partitioned as to make it "good" for *Allocator, Replicator, LoadBalancer* and *ParallelQueryProcessor*. The design quality is finally measured by the unified cost model. In other words, each potential fragmentation solution is tested for allocation, replication, load balancing, and query processing, respectively. The solution having the minimum cost is finally selected for the final PRDW design schema to be selected.

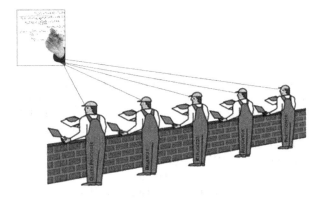

Fig. 2. Global approach to PRDW design

The main goal of this paper is to propose a cost model flexible enough to (*i*) provide a generic way for designing PRDW, by also exploiting probabilistic methods, (*ii*) reduce design cycle efforts, (*iii*) support of quick trade-off for evaluation purposes, (*iv*) integration of different PRDW design teams, (*v*) visibility and transparency of the different design phases. Indeed, coupling high flexibility and support for a wide number of functionalities as we argue in our proposal can easily introduce high complexity. To reduce this complexity, less-expensive algorithms are proposed and discussed in this paper.

The paradigm above converges in the PRDW design methodology $\mathcal{F}\&\mathcal{A}\&\mathcal{R}$, which follows and extends our previous proposals [6,10,11]. Summarizing, $\mathcal{F}\&\mathcal{A}\&\mathcal{R}$ is *a composite methodology where the main phases of PRDW design, i.e. Fragmentation, Allocation and Replication, are performed simultaneously instead that in an isolated manner*, like in traditional approaches. This conveys in a number of methodological and system-oriented advantages as well as practical achievements that we prove in this paper. Another relevant contribution of this research is represented by the special $\mathcal{F}\&\mathcal{A}\&\mathcal{R}$ query processing framework that defines *an innovative approach for effectively and efficiently supporting DW query processing (e.g., OLAP-style) over distributed fragments* that relies on a meaningful (query) cost model.

1.2   Paper Organization

The rest of the paper is organized as follows. In Sect. 2, we provide an overview of related work. Section 3 discusses four possible architectures for designing PRDW, highlighting benefits and limitations for each one. In Sect. 4, we provide a rigorous formalization of the PRDW problem according to our notation described in Sect. 1. Section 5 contains the details of the PRDW design methodology $\mathcal{F}\&\mathcal{A}\&\mathcal{R}$, which represents the main contribution of our research. In Sect. 6, we provide our query processing approach embedded in $\mathcal{F\&A\&R}$, along with its related cost model. Section 7 is focused on the experimental evaluation and analysis of $\mathcal{F\&A\&R}$, which clearly demonstrates the benefits deriving from our proposal. Finally, in Sect. 8, we derive conclusions of our research, and outline issues for future work.

2   Related Work

From active literature, we are not aware of any prior work that addresses the problem studied in this paper. That said, there is a tremendous amount of work, in both Parallel Data Warehouses and Replicated Data Partitioning. Both represent the background of our research, hence, in this Section, we survey relevant works in these areas. In more detail, PRDW design is related to four issues: data fragmentation, data allocation, data replication and load balancing.

Few studies on the issue of designing parallel data warehouses exist in literature. Noticeable ones are reported and discussed in the following.

2.1   Data Fragmentation

Fragmentation consists on the process oriented to decompose access objects (e.g., tables, materialized views, indexes) into a set of *disjoint partitions*. Fragmentation was introduced in the late 70s and early of 80s [16] as a *logic design technique* of traditional, distributed [28,49,51] and parallel [30] databases. With the development of Data Warehouses, fragmentation has become an *optimization problem focusing on efficiently representing and managing largest structures* in the context of physical design phase. In this Section, we outline main approaches in this scientific area.

Zilio *et al.* [64] focus on data partitioning issues of Shared-Nothing databases. More specifically, they addresses the problem of effectively selecting partitioning attributes, via proposing two specific algorithms: *Independent-Relation* and *Combined-Relation*, which differently exploit structural properties of attributes.

Stöhr *et al.* [58] propose an approach called *Multidimensional Hierarchical Fragmentation* for constructing and managing data warehouses on a disk-shared parallel machine having K disks. The fragmentation process consists in virtually partitioning each dimension table using the *interval mode* on attributes belonging to the lower levels of dimensional hierarchies, and consequentially partitioning the fact table on the basis of so-partitioned dimension tables. To speed-up queries, *bitmap join indexes* are selected via using attributes of dimension tables that belong to higher levels of dimensional hierarchies. To ensure a high parallelism degree and efficient load balancing, a *round-robin allocation* of fact fragments and associated bitmap indexes over the K disks is exploited.

Cuzzocrea *et al.* [23] propose an approach for fragmenting XML data warehouses characterized by three main steps: (i) extraction of selection predicates from the target query-workload; (ii) predicate clustering via the well-known

K-Means clustering algorithm; (*iii*) fragment construction on the basis of so-determined predicate clusters.

Boukhalfa *et al.* [8] propose a methodology and suitable algorithms (namely: genetic algorithm, Hill climbing, greedy algorithms) for partitioning relational data warehouses. Here, the fact table is partitioned on the basis of the fragmentation schemes of dimension tables. The criterion proposed to this end argues that each dimension table involved in any selection predicate is a candidate to referentially partition the fact table.

Bellatreche *et al.* [5] propose an innovative design methodology for *Shared-Nothing Data Warehouses* where the fragmentation and the allocation processes are done simultaneously. Here, the quality of the generated fragmentation schema is decided on the basis of the allocation process. Authors formalize the problem in terms of an optimization problem, and *genetic algorithms* are exploited to solve it. In [6,10,11], authors propose a comprehensive methodology for designing and querying PRDW over database clusters, called $\mathcal{F\&A}$ (the precursor of $\mathcal{F\&A\&R}$). $\mathcal{F\&A}$ assumes that cluster nodes are heterogeneous in processing power and storage capacity, and fragmentation and allocation phases are (again) performed in a simultaneous manner. Finally, in [6], the effectiveness of $\mathcal{F\&A}$ on a real-life parallel database machine (Teradata DBMS) is proven experimentally.

Nehme *et al.* [48] propose a tool that recommends the best partitioning configuration in distributed environments, given the input data warehouse and a reference query workload. The proposed tool recommends the set of tables to be partitioned in order to minimize the overall cost of the workload. Also, the proposed technique is embedded within the *parallel query optimizer* directly.

Pavlo *et al.* [53] propose a new approach for making automatic the partitioning of databases. To this end, they propose a cost model allowing us to estimate the analytical execution cost of an input query workload on the basis of the selected database partitioning schema. In more details, database design schema selection is based on the exploration of the target search space via graph theory. Experimental evaluation is provided by means of integrating the proposed approach into an in-memory DBMS.

2.2   Data Allocation

In actual literature, data placement proposal can be classified into two main classes: (1) papers where data allocation strategies are deployed on top of parallel platforms as round-robin, hash placement and range placement; (2) papers where data allocation strategies are based on attributes and deployed on top of distributed environments. Here, we describe the main works that have studied this issue.

Furtado *et al.* discuss partitioning strategies for *node-partitioned data warehouses*. In [33], authors propose a strategy that partitions the fact table on the basis of the larger dimension tables. Each large dimension table is hash-partitioned on its primary keys, while small dimension tables are replicated over all the processing nodes. The so-generated fragments are allocated in round-robin and random ways. In [34,35], authors exploit data replication for dynamic load-balancing while data availability is also considered as a constraint of the problem. Here, instead of node partitions, the target data set is divided into a much larger set of chunks and those are in turn divided into nodes in alternative ways that confer the desired load and availability balancing characteristics.

Karimi Adl et al. [41] use an ant colony optimization approach to solve the data allocation problem. Their goal is to design an effective data allocation schema which minimizes the total execution time of the workload and satisfied the storage constraint. Experimental results show that the proposed algorithm is capable of producing near-optimal solutions in a reasonable time.

Sarathy et al. [57] consider the fragment allocation problem as an integer non-linear problem with constraints, which has been already proved to be NP-hard. This formalization is further extended by [47] via adding more constraints as storage and power processing and, at the same time, achieving a simplified version of the general problem in terms of an integer programming problem.

Hababeh et al. [37] introduce a model that incorporates group sites in order to achieve high performance. The objective here is to minimize number of I/Os and communication cost between sites. A classification method is used to this end, which consists in grouping clustered sites, as to achieve the first goal of reducing the cost of communication between sites. After, availability and reliability are achieved due to the fact that multiple copies of the fragments are allocated.

Maik Thiele et al. [61] focus the attention on data allocation problems for real-time data warehouses. They consider a mixed workload (selection and update queries), and two specialized metrics: Quality of Service (QoS) and Quality of Data (QoD), respectively. The overall problem is formalized as a multi-objective problem, which has the characteristics of a backpack problem with additional inequality constraints. This so-formalized problem is finally solved by means of a linear programming algorithm.

2.3 Data Replication

Data replication is related to two major problems: (1) replica creation and (2) maintenance of materialized replicas. Our study focuses on the problem of creating replicas which is strongly influenced by the number of replicas and their placement. Indeed, the reference Replica placement Problem (RPP) consists in choosing the best replica placement on the distributed system in order to optimize given performance criteria. The optimal replica placement problem has been shown to be an NP-Hard problem [62]. Therefore, it is often solved by means of approximate solutions in a feasible time, and a relevant amount of work has been devoted to this paradigm in literature. Other solutions are based on hardware like RAID storage [15], but these are not considered in this research.

On the parallel architectures, Borr [14] proposes a naive approach, called Mirroring Declustering (MD), which maintains two copies of the same data, a primary and a backup copy, on two separate nodes, and tries to obtain a "good" load balancing. Along this line of research, Interleaved Declustering (ID) [60] argues to split the secondary copy into several partitions that are stored on disks. A further, interesting solution is represented by the an improved version called *Chained Declustering* (CD), by Hsiao *et al.* [39], according to which both the primary and the backup copy are stored on two adjacent nodes.

Finally, as regards distributed platforms, greedy algorithms [17,63], meta-heuristics [32] and data mining techniques [20] have been proposed in order to find the optimal placement of replicas or identify the optimal number of replicas [45] as well.

2.4   Load Balancing

Akal *et al.* [3] discuss *OLAP parallel processing* in a database cluster. They propose a virtual partitioning approach, called *Simple Virtual Partitioning* (SVP), which consists in fully replicating a database along a set of sites and breaking each query in sub-queries by adding suitable range predicates. Each node receives a sub-query and is forced to execute over a subset of fragments of the virtual partition.

Lima *et al.* [42] improve the SVP approach by addressing the issue of effectively determining the partition size. They introduce a novel approach, called *Adaptive Virtual Partitioning* (AVP), which dynamically tunes partition size. AVP starts by producing suitable sub-queries via adding range predicates over the virtual partitioning attributes. Each cluster node receives the *same* parameterized query plus a different dimensional range to be processed, and splits the range into sub-ranges assigned to sub-queries for processing purposes. Then, AVP tries to raise the range size until there is no performance degradation. Such proposal is further expanded in [43], where *partial replication* is used instead then full database replication.

Phan *et al.* [54] propose a framework for *coordination and optimization of OLAP processing performance on a cluster database*. The main objective here is to find the optimal load distribution that minimizes the cost of constructing materialized query tables, and the execution time associated to the sub-workload of each processing node. However, this solution is complex because it involves time to build materialized query tables, so that it impacts on the overall query execution. Indeed, the construction of such tables must consider storage space allocated to them, the computational power of processing nodes and the size of the space of all possible combinations of configurations. This problem has been formalized as a *combinatorial problem* whose search space is exponential in the number of queries, materialized query tables and processing nodes. In the authors' solution, a genetic algorithm is exploited to select the best matches among queries and processing nodes, and among materialized query tables and processing nodes, respectively.

Märtens *et al.* [46] focus the attention on *load balancing strategies for parallel processing of star schemas* equipped by bitmap join indices and deployed on Shared Disk (SD) architecture. In particular, an integrated scheduling strategy that simultaneously considers both processors and disks, regarding not only the total workload on each resource but also the distribution of load over time, is proposed. Gorla *et al.* [36] formalize the query allocation problem as a *multi-objective optimization problem*, and make use of a genetic algorithm to solve it.

Another line of research for improving the efficiency of load balancing is represented by *approximation/compression paradigms* (e.g., [21, 26, 27]). In fact, compressing replicas in some nodes may improve the load distribution as well as query efficiency.

To summarize, by analyzing research efforts done in the context of PRDW design, it emerges that *most state-of-the-art approaches are mainly concentrated on partitioning, allocation, replication and load balancing phases done iteratively and they use simple cost models that ignore guidelines related to the data skew and processing skew. In the area of replicated data partitioning, most existing works have mainly used sequential placement as main approach, and they haven't used any specific cost model to find the optimal replica placement.* Data replication is used only to improve the query processing over so-designed PRDW, and interaction between data replication and data placement (i.e., data fragmentation and data allocation) is not considered, apart from the main $\mathcal{F\&A}$ approach that originates $\mathcal{F\&A\&R}$. Figure 3 provides a summary of main issues and related solutions to the PRDW design problem.

3   Alternative Architectures for Designing PRDW

In this Section, we describe and critically discuss four possible alternatives for designing PRDW, by also highlighting benefits and limitations for each one. First, Fig. 4 sketches a generic PRDW design methodology. Here, input to the target PRDW design methodology consists of the following components: (i) the data warehouse schema $\mathcal{DWS}$; (ii) suitable statistics on the target data warehouse (e.g., relation cardinalities, column cardinalities, and skew information of columns); (iii) system information modeled in terms of: number of nodes M, *Processing Power* P_i of the node N_i, *Storage Capacity* S_i of the node N_i; (iv) workload of input OLAP queries (e.g., star queries) $\mathcal{Q} = \{Q_1, \ldots, Q_L\}$ and, for each query Q_i in $\mathcal{Q}$, its occurrence frequency f_i; (v) other important thresholds that provide the PRDW design methodology with additional information that helps into the decision making process, such as: *fragmentation threshold, replication threshold, system skew threshold.*

The generic PRDW design methodology above outputs the so-called *data placement schema*, which includes: (1) the *fragmentation schema* represented by a set of *partitioning attributes*; (2) the *allocation schema* represented by assignment of fragments over nodes (the allocation schema may either be redundant). The selected data placement schema is in charge of optimizing the execution cost of the workload $\mathcal{Q}$ and balancing its cost with respect to the benefit it provides. Therefore, designing a parallel shared-nothing data warehouse consists of three main steps: (i) fragmenting the data warehouse schema; (ii) allocating the generated fragments; (iii) developing effective and efficient query processing and optimization techniques.

Based on the challenges discussed above, it follows that our investigated PRDW design problem leads to three main issues, namely *data partitioning, fragment allocation* and *fragment replication*. Each one of the problems associated with issues above is known to be *NP-complete* [4, 9, 62]. In order to deal

Fig. 3. Issues and related solutions to the PRDW design problem

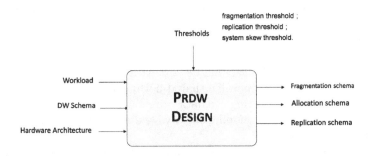

Fig. 4. Generic PRDW design methodology

with such a PRDW design problem, two main classes of methodologies are possible: *iterative design methodologies* and *combined design methodologies* [6,10,11]. From this first classification, four possible alternative architectures derive, which are depicted in Fig. 5.

Figure 5(*a*) shows an architecture where the basic idea consists in first fragmenting the RDW using any partitioning algorithm (e.g., [7,33,48,55,56,58,65]), then allocating the so-generated fragments by means of any allocation algorithm (e.g., [2,4,47]), and finally determining how to allocate the replicated fragments using any replication algorithm (e.g., [18,38,39,43,44]). According to this architecture, each partitioning, allocation and replication algorithm has its own cost model. The main advantage coming from this traditional methodology is represented by the fact that they are straightforwardly suitable to a large number of even-heterogeneous parallel and distributed environments (e.g., *Peer-to-Peer Databases*). Contrary to this, their main limitation is represented by the fact that they neglect the inter-dependency between the data partitioning, the fragment allocation phase and data replication phase, respectively.

Figure 5(*b*) depicts an architecture where the basic idea consists in first partitioning the RDW using any partitioning algorithm and then determining how fragments are allocated to the nodes by also determining replication of fragments. The main advantage of this architecture is that it takes into account the inter-dependency between allocation and replication, which are closely related [6,10,11]. Their main limitation is instead represented by the fact that they neglect the inter-dependency between the data partitioning and the fragment allocation phase, respectively.

Figure 5(*c*) shows an architecture such that the RDW is first horizontally partitioned into fragments, and then fragments are allocated to nodes within the *same* phase as in [5,19,56]. After that, a replication algorithm is exploited in order to determine how to allocate replicated fragments. The advantage of this architecture consists in performing the allocation phase at fragmentation time, in a simultaneous manner. However, the drawback of the architecture is that it neglects of the closely-related dependency between the allocation and the replication process.

Finally, Fig. 5(*d*) depicts an architecture such that fragmentation, allocation and replication are combined into a *unified process*. This architecture is suitable for designing PRDW from the scratch. In this context, three main issues play a critical role: *Data Partitioning* (DP), *Data Allocation* (DA) and *Data Replication* (DR). They are also closely inter-related among themselves as well. Discerning between fragmentation and allocation issues is, indeed, conceptually relevant. This because fragmentation issues deal with "local criteria" motivated by the same, native requirement of fragmenting data warehouses for efficiency purposes, while allocation issues deal with "physical placement" deriving from so-generated fragments over nodes. However, this difference must be introduced with extreme care as, generally speaking, it is not possible to determine the optimal fragmentation and allocation schemes by solving the two problems independently, since

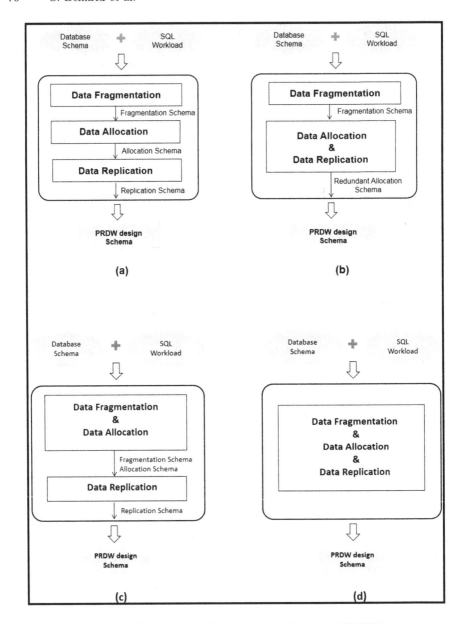

Fig. 5. Alternative architectures for designing PRDW

they are strongly inter-related. The data allocation phase, indeed, is in charge of deciding whether fragments will be replicated or not.

The basic difference among the four reference architectures depicted in Fig. 5 is indeed represented by the selection of partitioning attributes. The iterative approach determines the partitioning attributes using a cost model that neglects the inter-dependency between partitioning, allocation and replication. As a solid alternative, the joint approach decides the quality of the generated fragmentation scheme on the basis of its allocation process. The allocation process itself is combined with the replication. In other words, at partitioning time, a decision on the actual quality of the allocation scheme is taken, yet being this scheme susceptible of further changes, if needed.

4   PRDW Global Design Problem Formulation

In this Section, we formally introduce the PRDW design problem according to the notion of *global design* described in Sect. 1, such that all the aspects (i.e., partitioning, allocation, replication, load balancing and query optimization) are considered simultaneously within the main design phase. First, we provide the background formal definitions (Sect. 4.1) and, then, the formal problem definition (Sect. 4.2).

4.1   Formal Background

In the following, we provide formal definitions of main concepts and constructs to be used in our research.

1. *Attribute Skew Degree.* Given an attribute $\mathcal{A}$ such that its range of values is partitioned into $\mathcal{S}$ sub-domains ($\mathcal{S} > 1$), the *attribute skew degree* of $\mathcal{A}$, denoted by $\mathcal{SVA}(\mathcal{A})$, represents the standard deviation of the distribution of values (of $\mathcal{A}$) among the $\mathcal{S}$ sub-domains. Let $SEL(SD_i)$ denote the selectivity factor of a sub-domain SD_i, $\mathcal{SVA}(\mathcal{A})$ is defined as follows:

$$\mathcal{SVA}(\mathcal{A}) = \sqrt{\frac{1}{S} \times \sum_{i=1}^{S} \left(SEL(SD_i) - \frac{1}{S} \right)^2} \qquad (1)$$

It should be noted that our attribute skew degree is defined on top of the popular *standard deviation*, which measures how data of a given data distribution are distributed around the mean value. Therefore, the standard deviation indicates *the degree of consistency* among data. A big standard deviation means that there is more heterogeneity in the target data set. Conversely, a more homogeneous data set, consisting of data relatively close to the average value, exposes a smaller standard deviation value. This is a critical aspect in our approach, as we require a complete understating of properties of underlying data sets. To give an example on the attribute skew degree, let us consider Fig. 6, where the table *Customer* is partitioned into three distinct fragments, namely $Customer_1$, $Customer_2$ and $Customer_3$, via the attribute *City* and according to three disjoint sub-domains: $Dom(City) = d_1 \cup d_2 \cup d_3$ with $d_1 = \{Alger\}$, $d_2 = \{Rome\}$ and $d_3 = \{Paris\}$. The selectivity factor of there sub-domains is equal to 0.45, 0.1 and 0.45, respectively. By applying the formula (1), we obtain: $\mathcal{SVA}(City) = 12\%$.

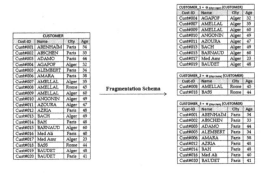

Fig. 6. Fragmentation schema for the table *Customer* of the running example

2. *Node Size.* The *size* of a node N_j, denoted by $Size(N_j)$, represents the sum of the fragment size allocated on N_j. Let $Size(F_i)$ denote the number of pages of the fragment F_i and $isStored(F_i, N_j)$ a boolean function that returns 1 if F_i is allocated on N_j, 0 otherwise, $Size(F_i)$ is defined as follows:

$$Size(N_j) = \sum_{i=1}^{NF} Size(F_i) \times isStored(F_i, N_j) \tag{2}$$

Focus again on the previous example of Fig. 6. Let us suppose that the fragment $Customer_1$ is allocated in the node N_1 and the remaining fragments (i.e., $Customer_2$ and $Customer_3$) are allocated in the node N_2. As a consequence, by applying the formula (2), we obtain: $Size(N_1) = 9$, $Size(N_2) = 11$.

3. *Node Load.* The *load* of a node N_j deriving from the evaluation of a star join query Q_k on N_j, denoted by $Load(N_j, Q_k)$, is the number of fragment tuples processed on N_j to evaluate Q_k. Formally:

$$Load(N_j, Q_k) = \sum_{i=1}^{NF} Size(F_i) \times isValid(F_i, N_j, Q_k) \tag{3}$$

such that $isValid(F_i, N_j, Q_k)$ is a boolean function that returns 1 if F_i is allocated on N_j and used for the evaluation of Q_k, 0 otherwise.

To give an example, consider again on the previous example of Fig. 6, and the following query Q:

```
SELECT COUNT(*)
FROM Customer
WHERE City = 'Alger' or City = 'Rome'
GROUP BY Cust_id
```

The rewriting of this query gives two sub-queries SQ_1 and SQ_2 whose syntax is as follows. For SQ_1:

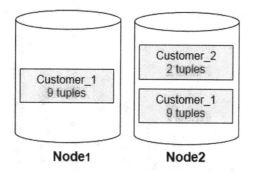

Fig. 7. Allocation schema for the table *Customer* of the running example

```
SELECT COUNT(*) FROM Customer_1 GROUP BY Cust_id
```

and, for SQ_1:

```
SELECT COUNT(*) FROM Customer_2 GROUP BY Cust_id
```

respectively. In particular, Q can be obtained from SQ_1 and SQ_2 as follows:

```
SELECT COUNT(*) FROM Customer_1 GROUP BY Cust_id
UNION
SELECT COUNT(*) FROM Customer_2 GROUP BY Cust_id
```

According to the allocation scheme shown in Fig. 7, sub-query SQ_1 is executed on the node N_1 and sub-query SQ_2 on the node N_2. As a consequence, the load of the node N_1 is $Load(N_1, SQ_1) = 9$ and the load of the node N_2 is $Load(N_2, SQ_2) = 2$.

4. *Mean Node Size.* Given a data warehouse $\mathcal{DW}$ partitioned into NF fragments $\mathcal{F} = \{F_1, F_2, \ldots, F_{NF}\}$ stored over M nodes, and the data placement skew factor[3] θ, the *Mean Node Size*, denoted by MPS, is defined as follows [59]:

$$MPS = \frac{1}{\sum_{j=1}^{M} \frac{1}{j^\theta}} \times \sum_{i=1}^{NF} Size(F_i) \qquad (4)$$

Considering again the running example, if the data placement skew equals 0.6, we must place at least 12 tuples at each node. Figure 8 shows the corresponding MPS for each data placement skew degree.

5. *Replication Degree.* The *degree of replication*, denoted by $\mathcal{R} = \{1, \ldots, M\}$ models the $\mathcal{R}$ physical copies of a fragment F allocated over the M nodes. In particular, $\mathcal{R} = 1$ models no replication, $\mathcal{R} = \mathcal{M}$ models full replication, and $1 < \mathcal{R} < M$ models the partial replication of *every* fragment for $\mathcal{R}$ times. For the partial replication, the degree of replication may be expressed in terms of percentage value, as follows:

---

[3] In this article, we suppose that the data skew adopt a *Zipf distribution* model.

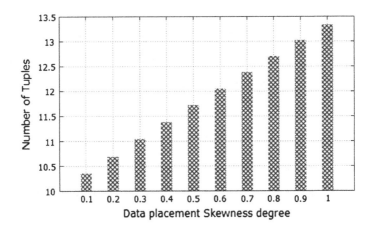

Fig. 8. Mean node size for the running example

$$\forall 1 < \mathcal{R} < M : \mathcal{CR}(\%) = \frac{(R-1) \times 100}{M} \tag{5}$$

6. *Load Balancing Skew Degree.* Given data warehouse $\mathcal{DW}$ partitioned into
   NF fragments $\mathcal{F} = \{F_1, F_2, \ldots, F_{NF}\}$ stored over the M nodes, we say that
   a system is balanced if the distance between all the loads and the gravity
   center[4] is null. To obtain the value between 0 and 1, we use the *Normalized
   Euclidean Distance*. Therefore, the *Load Balancing Skew Degree* of a star join
   query Q_j, denoted by $LDLB(Q_j)$, is defined as follows:

$$LBDS(Q_j) = \sqrt{\sum_{i=1}^{M} \frac{(Load(N_i, Q_j) - MeanLoad)^2}{\sigma^2}} \tag{6}$$

where $MeanLoad = (\frac{1}{M} \sum_{i=1}^{M} Load(N_i, Q_j))$ and σ is the standard deviation.
By applying the formula 6 on the running example, we obtain: $LBDS(Q) = 20\%$.

4.2   Formal Problem Definition

In our global PRDW design approach, the fragmentation process is the core of
the PRDW design methodology, $\mathcal{F}\&\mathcal{A}\&\mathcal{R}$, and, consequentially, the quality of
the PRDW design methodology itself strongly depends on the quality of the
fragmentation process. As a consequence, we connote the $\mathcal{F}\&\mathcal{A}\&\mathcal{R}$ methodology as a *fragment-driven PRDW design methodology*. Formally, the subtended
fragment-driven PRDW design problem, which globally considers the different

---

[4] The gravity center represents the point around which the mass is distributed sym-
metrically. In our study, the gravity center represents the average of loads.

aspects of the PRDW design problem (i.e., partitioning, allocation, replication, load balancing and query optimization) can be formalized in terms of a *Constraint Optimization Problem*, as follows.

Formally, given:

- a database cluster machine $\mathcal{DBC}$ with M nodes $\mathcal{N} = \{N_1, N_2, \ldots, N_M\}$;
- a relational data warehouse $\mathcal{RDW}$ modeled according to a star schema and composed by one fact table $\mathcal{F}$ and d dimensional tables $\mathcal{D} = \{D_1, D_2, \ldots, D_d\}$ – similarly to [43], we suppose that all dimensional tables are replicated over the nodes of the database cluster and are in their main memory;
- a set of star join queries $\mathcal{Q} = \{Q_1, Q_2, \ldots, Q_L\}$ to be executed over $\mathcal{DBC}$, being each query Q_l characterized by an access frequency f_l;
- the *maintenance constraint* $\mathcal{W}$, such that $\mathcal{W} > \mathcal{N}$, representing the number of fragments that the designer considers relevant for his/her target allocation process (note that this number must be greater than the number of nodes, i.e. $\mathcal{W} \gg M$);
- the *replication constraint* $\mathcal{R}$, such that $\mathcal{R} \leq \mathcal{M}$, representing the number of fragment copies that the designer considers relevant for his/her parallel query processing;
- the *attribute skewness constraint* θ representing the degree of non-Uniform value distributions of the attribute sub-domain chosen by the designer for the selection of the fragmentation attributes;
- the *data placement constraint* α representing the degree of data placement skew that the designer allows for the placement of data;
- the *load balancing constraint* δ representing the data processing skew that the designer considers relevant for his/her target query processing;

the problem of designing a PRDW from $\mathcal{DWS}$ over the database cluster $\mathcal{DBC}$ consists in *fragmenting the fact table $\mathcal{F}$ into $\mathcal{NF}$ fragments and allocating them and the replicated fragments over different $\mathcal{DBC}$ nodes such that the total cost of executing all the queries in $\mathcal{Q}$ can be minimized while all constraints of the problem are satisfied.*

5 $\mathcal{F\&A\&R}$: A Novel Methodology for Designing PRDW According to a Global Approach

In this Section, we describe in detail our proposed PRDW design methodology, $\mathcal{F\&A\&R}$, which follows and extends our previous proposal [6,10,11]. As highlighted in Sect. 1.1, the main idea of $\mathcal{F\&A\&R}$ consists in performing the main phases of PRDW design (i.e., fragmentation, allocation, and replication) simultaneously. Figure 9 sketches the flowchart of $\mathcal{F\&A\&R}$.

In the following, we describe in details the two main phases of $\mathcal{F\&A\&R}$, i.e. partitioning and allocation. It should be noted here that replication is directly related to the allocation phase, being these two latter phase performed within the same main process (i.e., allocation). Overall, all these phases encompass steps described by the flowchart in Fig. 9.

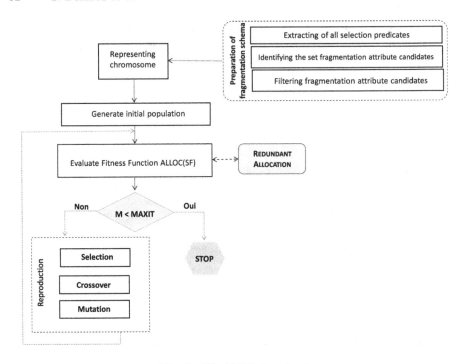

Fig. 9. $\mathcal{F}\&\mathcal{A}\&\mathcal{R}$'s flowchart.

5.1 $\mathcal{F}\&\mathcal{A}\&\mathcal{R}$ Partitioning Phase

To select a suitable horizontal partitioning schema, we adapt our genetic algorithm proposed in [5]. Representing chromosomes that model candidate fragmentation schemas is the most probing tasks when applying genetic algorithms to the
PRDW design problem. Each chromosome may be represented as a *multidimensional array* that models the partitioning domain of a fragmentation attribute.
To identify the partitioning attribute candidate, we perform the following tasks.
(1) Extract *all* selection predicates exploited by the input queries. (2) Assign to
each dimension table D_i, such that $1 \leq i \leq d$, the set of selection predicates they
are involved to, denoted by $SSPD_i$. (3) Ignore dimension tables D_i having an
empty set $SSPD_i$ (i.e., they will not participate in the fact table fragmentation
process). (4) Identify the set of fragmentation attribute candidates. (5) Eliminate attributes having high skew and that do not satisfy the attribute skew
constraint. (6) Decompose domain values of each fragmentation attribute into
sub-domains (each sub-domain may be represented by a simple predicate along
with its selectivity factor defined on the fact table).

Once the set of fragmentation attributes is selected, our proposed genetic
algorithm generates a random population that contains several chromosomes. For
each chromosome, our algorithm checks if it satisfies the following maintenance
constraint:

$$NF_i \leq W \tag{7}$$

such that NF_i represents the number of fragments on the node N_i. If it is the case, the associated chromosome is kept in the population; otherwise, *merges operations* are applied with the goal of reducing its number of fragments. Once initial population is created, our genetic algorithm performs further operations, such as *crossover* and *mutation*, in order to improve the quality of the actual population. The application of these operators is monitored by an ad-hoc *evaluation function* that allocates the so-generated fragments of each valid chromosome over the nodes of the target parallel machine. The allocation phase is described in details in Sect. 5.2. Once allocation has been performed, the cost of executing queries over nodes is estimated, and, the chromosome population that allows the minimum query cost is finally selected as the reference fragmentation schema.

5.2 $\mathcal{F}\&\mathcal{A}\&\mathcal{R}$ Allocation Phase

The data allocation problem consists in determining the best placement of a set of fragments over database cluster nodes with the goal of minimizing the cost of evaluating queries belonging to a given (query) workload $\mathcal{Q}$. In distributed and parallel databases, as well as data warehouses, this problem can be formalized in terms of a *clustering problem*. This because clustering means to "place" a set of entities into a given number of groups according to a given measure of their tendency to be used together, yet conveying in a conceptual basis that is similar to the one of the problem we investigate. This is illustrated by Fig. 10.

The fragment allocation process is closely related to the fragment replication problem. In other words, the data allocation algorithm is in charge of deciding whether fragments will be replicated or not. To this end, we propose using a *fuzzy clustering method*, namely the *fuzzy k-means clustering algorithm* [13]. In fuzzy clustering techniques, data points can belong to more than one cluster, and associated with each of the points are so-called *membership degrees*, which represent the degree at which data points belong to the different clusters. The underlying principle in fuzzy clustering is represented by the criterion of assigning data elements to multiple clusters, with varying degree of membership. In particular, membership degrees between 0 and 1 are used instead of crisp assignments of data in clusters. Fuzzy clustering is often better suited than classical clustering techniques as there is often no sharp boundaries among clusters of data.

Based on this main evidence, in the context of the $\mathcal{F}\&\mathcal{A}\&\mathcal{R}$ methodology, we propose a new allocation procedure that is based on *fuzzy clustering of fragments*. Let us formulate the deriving fragment allocation problem as follows. Consider a set of fragments $\mathcal{F} = \{F_1, F_2, \ldots, F_{NF}\}$ with dimensions in the Euclidean space $\Re^d$ (i.e. $F_j \in \Re^d$). The problem of fragment allocation via fuzzy clustering consists is performing a partitioning of these fragments into M fuzzy sets with respect to a given criterion, being M the number of $\mathcal{DBC}$ nodes (see Sect. 4.2). The criterion is usually determined in terms of the optimization of a suitable

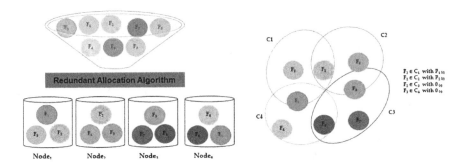

Fig. 10. Conceptual similarity between allocation and clustering

objective function. The result of the fuzzy clustering can be expressed as a partitioning matrix U, such that $U = [i][j] = u_{ij}$, $i = 1..M$ and $j = 1..NF$, where u_{ij} is a value in $\{0, 1\}$ which expresses the membership degree of the actual fragment. Also, a further constraint states the total membership degrees of a given fragment $F_j \in \mathcal{F}$ in all classes must be equal to 1, for all the nodes $N_i \in \mathcal{N}$, with $i = 1..M$, and all the fragments $F_j \in \mathcal{F}$, with $j = 1..NF$, i.e.:

$$\sum_{i=1}^{M} u_{ij} = 1, \quad \forall j = 1..NF \tag{8}$$

The objective function f_O to be *minimized* is defined as follows:

$$f_O = \sum_{k=1}^{NF} \sum_{i=1}^{M} u_{ij}^{m} \, ||X_k - V_i||^2 \tag{9}$$

wherein: (i) $m > 1$ models a *degree of fuzziness* that governs the influence of membership degrees; (ii) X_k models a vector of data points; (iii) V_i models the center of cluster C_i; (iv) $||X_k - V_i||^2$ models the Euclidean distance between X_k and V_i.

Given the formal problem definition above, the $\mathcal{F\&A\&R}$ allocation phase consists of the following steps:

1. *construction of the Fragment Usage Matrix (FUM);*
2. *fragment representation in $\Re^2$;*
3. *construction of the Fragment Membership Matrix (FMM);*
4. *fragment clustering;*
5. *construction of Fragment Placement Matrix (FPM).*

The final FPM, of course, provides us with the solution of the allocation problem.

In the following Sections, we describe the previously-listed steps of the $\mathcal{F\&A\&R}$ allocation phase.

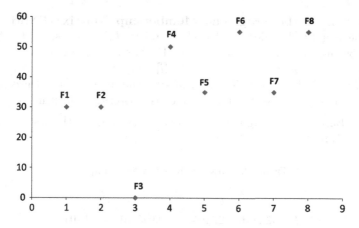

Fig. 11. Fragment representation associated to the FUM of Table 1

Construction of the Fragment Usage Matrix (FUM). FUM models the usage of fragments according to the set of queries in $\mathcal{Q}$. FUM contains queries as rows and fragments as columns. The value $FUM[i][j]$, such that $1 \le i \le L$ and $1 \le j \le NF$, is equal to 1 if the query Q_i involves the fragment F_j; otherwise, it is equal to 0. An additional column is added to represent the *access frequency* f of each query.

Example 1: Let $F = \{F_1, F_2, F_3, F_4, F_5, F_6, F_7, F_8\}$ and $Q = \{Q_1, Q_2, Q_3, Q_4\}$ be the set of so-generated fragments and queries, respectively. A possible FUM is shown in Table 1. □

Table 1. FUM of the running example

Queries	F_1	F_2	F_3	F_4	F_5	F_6	F_7	F_8	f
Q_1	1	1	1	0	1	0	1	0	20
Q_2	1	1	1	1	0	0	0	0	35
Q_3	0	0	1	0	1	1	1	1	30
Q_4	1	1	1	1	1	1	1	1	15

Fragment Representation in $\Re^2$. Each fragment F_i is represented in the two-dimensional space $\Re^2$ by suitable coordinates (x, y). Given a fragment F_i in $\Re^2$, these coordinates are based on the frequency of queries that do not involve the fragment F_i.

Example 2: Figure 11 shows the fragment representation associated to the FUM of Table 1. □

Construction of the Fragment Membership Matrix (FMM). FMM models the membership degree of each fragment F_i with respect to the cluster C_i according to the set of queries in $\mathcal{Q}$. FMM contains fragments as columns and clusters as rows. The value $FMM[j][i]$, such that $0 \leq i \leq NF - 1$ and $0 \leq j \leq M - 1$, ranges over $[0, 1]$ and models the membership degree of F_i to C_i, computed according to the fuzzy k-means clustering algorithm [13].

Example 3: Based on the fragment representation of Fig. 11, the associated FMM is shown in Table 2. □

Table 2. FMM of the running example

	C_0	C_1	C_2	C_3
F_0	5,01E-03	2,79E-04	9,94E-01	4,35E-04
F_1	5,92E-03	2,72E-04	9,93E-01	4,29E-04
F_3	3,56E-09	1,00E+00	4,87E-09	1,53E-09
F_4	6,66E-02	6,07E-03	3,75E-02	8,90E-01
F_5	9,70E-01	7,97E-04	2,64E-02	2,84E-03
F_6	5,28E-03	6,94E-04	3,27E-03	9,91E-01
F_7	9,79E-01	7,83E-04	1,77E-02	2,82E-03
F_8	1,37E-02	1,81E-03	8,28E-03	9,76E-01

Fragment Clustering. In order to generate groups of fragments into clusters, we make use of the basic principle claiming that *larger membership degrees indicate higher confidence in the assignment of objects to the actual cluster.* On the basis of this main insight, we sort membership degrees in descending order and we assign the fragment F_i to the $\mathcal{R}$ first clusters, being $\mathcal{R}$ the replication degree (see Sect. 4.1), such as the data placement constraint α (see Sect. 4.2) is satisfied. This step generates a set of clusters $\mathcal{C} = C_0, \ldots, C_{M-1}$, such that each cluster C_i represents a sub-set of fragments.

Example 4: Figure 12 depicts the fragment clustering associated to the FMM of Table 2. □

Construction of Fragment Placement Matrix (FPM). FPM models the positions of a fragment across nodes (recall that fragment replicas may exist). To this end, FPM rows model fragments, whereas FPM columns model nodes. $FPM[i][m] = 1$, with $1 \leq i \leq NF$ and $1 \leq m \leq M$, if the fragment F_i is allocated on the node N_m in $\mathcal{N}$, otherwise $FPM[i][m] = 0$. Our allocation procedure considers clusters as "movable units" during allocation: clusters are placed in a round-robin fashion over nodes.

Example 5: Clusters generated from the fragment clustering of Fig. 12 are placed in a round-robin fashion over processing nodes: the associated FPM is shown in Table 3. □

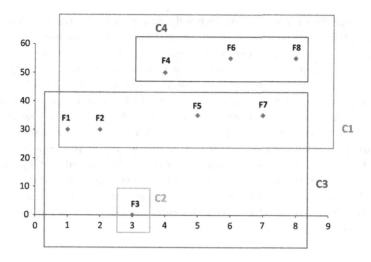

Fig. 12. Fragment clustering associated to the FMM of Table 2

Table 3. FPM of the running example

	F_1	F_2	F_3	F_4	F_5	F_6	F_7	F_8
N_1	1	1	0	1	1	1	1	1
N_2	0	0	1	0	0	0	0	0
N_3	1	1	1	0	1	0	1	0
N_4	0	0	0	1	0	1	0	1

6 $\mathcal{F}\&\mathcal{A}\&\mathcal{R}$ Query Processing Framework

In this Section, we provide the details on the $\mathcal{F}\&\mathcal{A}\&\mathcal{R}$ query processing framework that, as highlighted in Sect. 1, relies on an innovative (query) cost model.

First, focus the attention on the query mechanism supported by a conventional PRDW. Once the fragmentation schema is generated and the so-generated fragments are placed, *global queries* posed to the data warehouse are then rewritten over fragments and evaluated on the database cluster $\mathcal{DBC}$. The ideal parallel query processing method optimizes a smaller set of queries and tries to minimize the total execution cost for the entire set of queries. To evaluate a given query, first valid fragments and their locations across nodes should be identified.

To this end, $\mathcal{DBC}$ contains multiple *Processing Nodes* (PN), which are responsible for processing only the data warehouse rows on its own disks. Each PN node is modeled as a single CPU equipped with disks and a buffer pool. The execution of a query is managed by specialized nodes, called *Coordinator Nodes* (CN), which are devoted to the following tasks: (*i*) re-writing queries, (*ii*) scheduling queries, (*iii*) merging results and (*iv*) reporting them back to the client.

The evaluation of a query can be briefly summarized by the following steps. (1) When a CN node receives a query from the user, it re-formulates the query and converts the query into a set of sub-queries such that each one executes on a singleton fragment. The sub-queries are then added to the workload queue. (2) The scheduler maintains a queue of queries that are to be evaluated. To this end, the scheduler analyzes data requirements for the evaluation of queries and determines a favorable assignment of queries to a PN node. (3) The PN node passes queries to its data warehouse instance for evaluation. Answers to these queries are asynchronously sent back to CN nodes. These queries are marked as processed and removed from the workload queue. (4) Once all sub-queries have been processed and the intermediate results submitted to CN nodes, CN nodes merge different results and perform every appropriate aggregation processing needed. (5) Finally, a suitable CN nodes composes the final result and delivers it to the user.

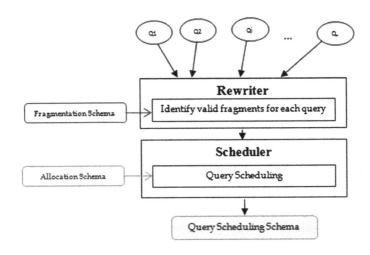

Fig. 13. Query scheduler conceptual architecture

From our query evaluation mechanism, it clearly follows that, since our allocation process is redundant (i.e., each fragment can have several placements by means of replicas), we make use of a suitable *Query Scheduler* to find the best allocation of each sub-queries. The conceptual architecture of our Query Scheduler is depicted in Fig. 13. It should be noted here that each valid fragment gives rise to a sub-query.

Formally, the query processing of $\mathcal{F}\&\mathcal{A}\&\mathcal{R}$ can be formalized as follows. Given:

– a set of fragments $\mathcal{F} = \{F_1, F_2, \ldots, F_{NF}\}$, being each fragment F_i, with $1 \leq i \leq NF$, characterized by its size $Size(F_i)$;
– a database cluster machine $\mathcal{DBC}$ having $\mathcal{M}$ nodes $\mathcal{N} = \{N_1, N_2, \ldots, N_M\}$;

- a set of star queries $\mathcal{Q} = \{Q_1, Q_2, \ldots, Q_L\}$ to be executed over $\mathcal{DBC}$, being each query Q_l, with $0 \leq l \leq L-1$, characterized by an access frequency f_l;
- the *processing skew constraint* ϱ representing the data processing skew that the designer considers relevant for his/her target query allocation process;

determine the following state function:

$$isAllocated(Q_i, N_j) = \left\{ \begin{array}{l} 1 \; Q_i \; \; on \; \; N_j \\ 0 \; otherwise \end{array} \right.$$

such that Q_i denotes a query of the target query workload $\mathcal{Q}$ and N_j denotes a node of the node set $\mathcal{N}$, by minimizing the total query processing cost due to evaluating *all* the queries in $\mathcal{Q}$ while maximizing the productivity of each node in $\mathcal{N}$, subject to the fixed processing skew constraint ϱ.

The above-introduced query processing defines an NP-hard problem, which is similar to a *Dual Bin Packing Problem* (DBPP) [40]. To provide sub-optimal solutions to this problem, we propose a proper *greedy algorithm* that is in charge of executing the query scheduling for supporting star query evaluation against the parallel machine (see Algorithm 1).

Algorithm 1. Query Allocation(Q_j, $\mathcal{N}$)

1: Let `ListFrag` the list of valid fragments for Q_j.
2: Let `NumberFrag` the number of fragments in `ListFrag`;
3: Let `NumberValidNodes` the number of valid nodes for the fragments in `ListFrag`;
4: Let `ListSubQueries` the sub-query list (of Q_j); /**each valid fragment gives rise to a sub-query**/
5: Estimate the number of IOs needed to evaluate Q_j, $Size(Q_j)$;
6: Compute the MPS (4) of Q_j, as follows:

$$MPS = \frac{1}{\sum_{j=1}^{NumberValidNodes} \frac{1}{j^\delta}} \times Size(Q_j) \qquad (10)$$

7: Sort `ListFrag` according to their size in descending order;
8: **for** $i = 1$ to $NumberFrag$ **do**
9:    Get the valid nodes for the ith fragment in `ListFrag` and store them in the list `ListNode`;
10:    Compute the load of each node from `ListNode`;
11:    Assign F_i to the node with largest residual capacity;
12: **end for**

Focus the attention on Algorithm 1. First, we identify the valid fragments and their associated sub-queries, the number of valid fragments and the set of valid nodes needed for the evaluation of the sub-queries (of Q_j – lines 1–4). Next, we estimate the number of IOs needed to evaluate Q_j (lines 5) and we compute the processing bound MPS (line 6). We then sort valid fragments in descending order (line 7) and, for each so-generated sub-query (line 8), we perform the following steps: (1) select the valid nodes; (2) compute the load of each valid node; (3) pick the sub-query at the node having the largest residual capacity (lines 9–12). This finally realizes the scheduling of sub-queries on fragments and their replicas, so that giving the support for their evaluation.

Once the query allocation process has been performed, we compute the execution cost of $\mathcal{Q}$ over the $\mathcal{M}$ nodes of $\mathcal{DBC}$ in terms of number of IOs, according to the following equation:

$$\sum_{l=1}^{L} MAX_{1 \leq j \leq M} \left(\sum_{i=1}^{NF} MUF[i][k] \times MPF[i][j] \times Size(F_i) \right) \qquad (11)$$

7   Experimental Evaluation and Analysis

This Section reports on the results of the experimental evaluation of our proposed PRDW design methodology $\mathcal{F}\&\mathcal{A}\&\mathcal{R}$. To this end, we first describe the experimental framework, data sets and query workloads, and then we present the obtained results.

Our simulation conducted on a computer with $2.8\,GHz$ Intel Pentium Core Duo equipped with $3\,GB$ RAM. Algorithms were carried out in Java programming language. For the hardware architecture, we simulate a homogeneous database Cluster of 10–32 nodes.

In our experimental assessment, we use the dataset and queries defined in the Star Schema Benchmark SSB (Fig. 14) [50], which is a derivative of the benchmark TPC-$H2$ [52]. In particular, we generated several instances of the SSB benchmark data set by using the data generator supplied with SSB. The size of each instance is controlled by a *scale factor*, denoted by SF. A value $SF = X$ results in a data set of size XGB, with 94 % of the data stored in the fact table.

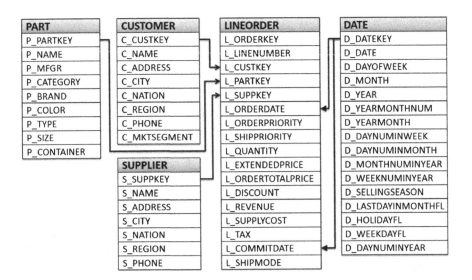

Fig. 14. *SSB* star schema

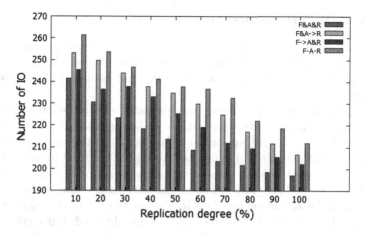

Fig. 15. Performance of $\mathcal{F\&A\&R}$ against comparison PRDW design approaches

We limited the maximum value of SF to 100 (i.e., a $100\,GB$ data set) in order to ensure the timely execution of the test workloads on our single experimental machine.

The full set of SSB queries consists of 13 queries. We generated workloads of star queries from the queries specified in the benchmark. We also excluded queries $Q_{1.1}$, $Q_{1.2}$, and $Q_{1.3}$ from the workload because of these queries contain selection predicates on fact table attributes, and this functionality is not yet supported by our prototype. This modification does not affect the generality and the reliability of the generated query workloads. In more detail, as regards experiments presented in this Section, we augmented the query workload to be 4 times larger. The so-obtained 36 queries are derived from the original 13 queries via varying the target predicate values. Specifically, we first convert each benchmark query to a template, by substituting each range predicate in the query with an abstract range predicate, e.g. $s\_region = \text{'}UNITED\ STATES\text{'}$ is converted into $s\_region = Reg$, where Reg is a parametric variable. To obtain a workload query, we simply substitute the abstract ranges in the query template with concrete predicates based on parameters that controls the selectivity of the query itself. We have used 20 selection predicates defined on 8 different attributes: $\{s\_region, d\_year, s\_nation,$ $c\_city, c\_region, s\_city, p\_category, c\_nation\}$. The domains of these attributes are split into: 7, 5, 7, 6, 5, 6, 3 and 8 sub-domains, respectively, to perform our proposed PRDW design approach.

We performed several kinds of experiments by varying several important factors such as replication degree, data partitioning skew, attribute skewness factor and data processing skew factor, to obtain a "rich" and reliable experimental evaluation of $\mathcal{F\&A\&R}$.

7.1 $\mathcal{F}\&\mathcal{A}\&\mathcal{R}$ Performance Analysis

As a first experiment, we study the performance of our proposed methodology $\mathcal{F}\&\mathcal{A}\&\mathcal{R}$. We conducted two kinds of experiments, we set the fragmentation threshold to 100; attribute skew was set to 0.5 and data partitioning skew has been neglected.

In the first kind of experiments, we compared our proposed methodology $\mathcal{F}\&\mathcal{A}\&\mathcal{R}$ against the three iterative PRDW design approaches: (1) partitioning, allocation and replication are treated in isolation, (2) partitioning and allocation are treated in joint manner and in isolation to replication, and (3) allocation and replication are traited in joint manner and in isolation to replication. For each PRDW design methodology, we measured the query execution time on a 10-node database cluster machine versus the variation of the replication degree $\mathcal{R}$ over the interval $[1:10]$. Figure 15 shows the results obtained and confirms to us that the combined approach outperforms the iterative one significantly. From derived results, we observe that an increase of the replication factor involves in an increase of the whole system performance. Also, we observe that increasing the replication factor involves the minimization of the system throughput, by balancing the load among the $\mathcal{DBC}$ nodes.

The computational overhead of the four approaches is depicted in Fig. 16. From Fig. 16, it clearly follows that $\mathcal{F}\&\mathcal{A}\&\mathcal{R}$ introduces a bigger computational overhead than other comparison approaches. Therefore, in future work the performance of actual algorithms must be improved.

The second kind of experiments focuses on checking whether the $\mathcal{F}\&\mathcal{A}\&\mathcal{R}$ parallel processing is characterized by a *linear speed-up*. To this end, we considered a 32-node database cluster machine and we engineered four different scenarios characterized by a proper replication factor $\mathcal{R}$: $8(25\%), 16(50\%), 24(75\%), 32(100\%)$. We ranged the node number factor from 1 to 32 and, for each value,

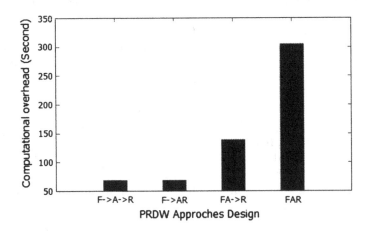

Fig. 16. Computational overhead performance of $\mathcal{F}\&\mathcal{A}\&\mathcal{R}$ against comparison PRDW design approaches

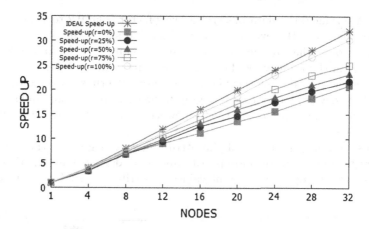

Fig. 17. Effect of replication factor on the linear speed-up of $\mathcal{F\&A\&R}$

we calculated the speed-up. The results shown in Fig. 17 confirm to us that an increase of the replication factor results in raising the speed-up and making it more linear. It should be noted that, in the case $\mathcal{R} = 100\,\%$, the speed-up is approximately linear. This is due to the fact that the load balancing resulting from the replication does not completely eliminate negative effect of data skews.

7.2   Inter-dependency Among PRDW Design Parameters

The second experiment that we conducted studies the dependence between the $\mathcal{F\&A\&R}$ parameters. In the first test, we studied the dependence between attribute skew and data partitioning. We fixed the fragmentation threshold to 100 and the node number to 10. We ranged the data attribute skew factor from 0.2 to 1 and, for each value, we calculated the data partitioning skew degree. Figure 18 shows the obtained results, and confirms that the data partitioning skew increases greatly when the attribute skew factor increases.

In the second assessment, we studied the effect of the replication degree on the parallel processing. We ranged the replication factor from 1 to 10 and, for each value, we calculated the processing skew degree. As shown in Fig. 19, the increase of the replication factor may almost completely eliminate negative effects of data skews with respect to the partitioning attributes. On the other hand, the efficiency of the PRDW increases greatly when the replication factor increases, as expected.

7.3   Effect of Attribute Skew Degree

The third experiment that we conducted studies the effect of data partitioning. We fixed the attribute skew to 0.5 and the fragmentation threshold to 100. Experiments were conducted with respect to four different scenarios characterized by a proper replication factor $\mathcal{R}$: 2, 5, 8, 10, on top of a 10-node database cluster machine. We ranged the data placement skew factor from 0.2 to 1 and, for each value, we calculated the execution time. Figure 20 shows the obtained results, and demonstrates that an increase of the partitioning skew factor degrades the performance of the framework. This is because the framework imbalance degree increases as the partitioning skew factor increases. We thus conclude that either partitioning skew or attribute skew degrades the performance of parallel processing, and that the performance of the framework increases greatly when the replication factor increases.

We also studied the effect of the attribute skew degree on the number of fragments generated and the data partitioning skew degree. This experiment was performed by setting three different values of attribute skew factor on top of a

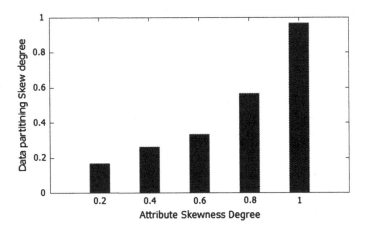

Fig. 18. Dependency between attribute skew and data partitioning skew

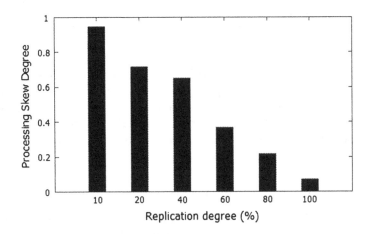

Fig. 19. Dependency between replication degree and processing skew

10-node database cluster machine, and by ranging the fragmentation threshold from 100 to 350. For each of these values, we calculated the execution time. Results are shown in Fig. 20. The obtained results show that the attribute skew degree limits the number of generated fragments because our proposed methodology eliminates attributes having a high skew degree from the list of partitioning attribute candidates. Although the number of fragments is small, the performance of the framework is good because the degree of load imbalance is also small. This is confirmed by the Fig. 21, where the (corresponding) variation of the query execution time is shown.

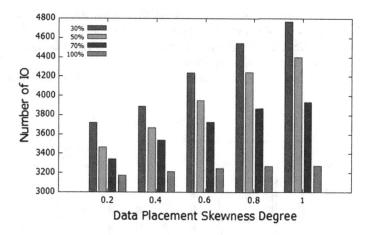

Fig. 20. Effect of attribute skew degree on query execution time

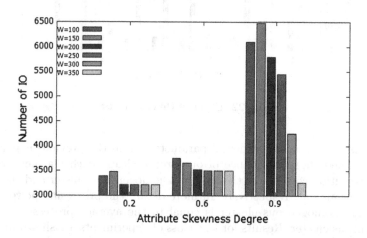

Fig. 21. Effect of attribute skew degree on fragmentation threshold

7.4   Effect of Heterogeneity

In previous experiments, we have assumed that the cluster is homogeneous (i.e., all nodes have the same processing power). Here, we investigate the effect of heterogeneity. We devised an experimental environment where processing power of each node have been generated according to a random distribution, thus obtaining a totally heterogeneous database cluster environment. With this novel experimental setting in mind, we first normalized the processing power of nodes, as to obtain a truly homogenous database cluster environment. After, we adapted our proposed PRDW design methodology as follows:

- *allocation algorithm*: we use our $\mathcal{F\&A}$-$\mathcal{ALLOC}$ algorithm [11], the core of $\mathcal{F\&A}$ [6,10,11], as to assign classes of fragments to nodes – it should be noted that algorithm $\mathcal{F\&A}$ -$\mathcal{ALLOC}$ runs on nodes characterized by heterogeneous processing power and storage capability;
- *query scheduling*: we assign each sub-queries to the most powerful node that can treat it.

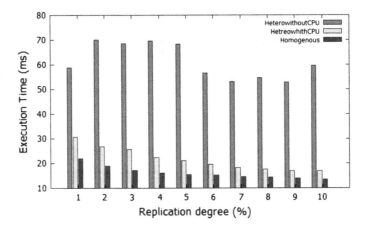

Fig. 22. Effect of Heterogeneity

We kept the same experimental parameters as in the previous experiments, and we studied the performance of our proposed approach via measuring the average execution cost due to the load of a homogeneous cluster and the load of a heterogeneous one, respectively. In more detail, the processing power of each node in the homogeneous cluster is equal to the average processing power of heterogeneous cluster. Results for this class of experiments are shown in Fig. 22, where the advantages that derive from taking into account the heterogeneity of cluster node characterise clearly emerges.

8   Conclusions and Future Work

It is well-known that, in order to obtain high performance with PRDW, it is critical to introduce a suitable design methodology as this allows us to ensure the effectiveness and efficiency of the system. Despite this, classical design approaches for the related context of parallel databases have not changed for years, whereas designing PRDW remains a difficult problem. Indeed, this problem includes a set of sub-problems: (*i*) the choice of hardware architecture, (*ii*) data fragmentation, (*iii*) data allocation, (*iv*) data replication, (*v*) load balancing and (*vi*) query processing. Each of these ones is known to be a NP-hard problem. Therefore, all state-of-the-art approaches are likely to have limitations and drawbacks. Indeed, these approaches do not consider the inter-dependency among sub-problems related to the design of PRDW, and they exploit heterogeneous metrics for assessing the "quality" of the final design (a different metrics for every different phase).

Contrary to this trend of classical approaches, in this paper we have proposed a novel design approach called $\mathcal{F\&A\&R}$, whose main benefit consists in performing the major PRDW design phases (i.e., fragmentation, allocation and replication) simultaneously. We demonstrated the advantage deriving from interpreting the PRDW design problem as an unified problem, and, as a consequence, we conferred to the fragmentation process the role of core method of the whole PRDW design methodology. To this end, we exploited a genetic algorithm for supporting the fragmentation process proposed by us in [5,6], plus an original redundant data allocation process that is based on fuzzy k-means clustering. Finally, our cost model, which significantly connotes our $\mathcal{F\&A\&R}$ query processing framework integrates the concepts of all the design phases, and the deriving query scheduling was formalized and solved in terms of a DBPP problem. As a secondary contribution of our research, we have evaluated our approach over the well-known SSB benchmark. Observed results are very promising.

Future work is oriented towards two different directions: (*i*) development of advanced algorithms capable of parallelizing the various steps of our PRDW design methodology; (*ii*) extending our query cost model by considering, in addition to the actual parameters, even the interaction among target queries; (*iii*) dealing with innovative characteristics of modern data warehouse applications, such as *visualization* (e.g., [24]) and *security* (e.g., [25]) issues.

References

1. Agrawal, D., Das, S., El Abbadi, A.: Data Management in the Cloud: Challenges and Opportunities. Synthesis Lectures on Data Management. Morgan & Claypool Publishers, San Rafael (2012)
2. Ahmad, I., Karlapalem, K., Ghafoor, R.A.: Evolutionary algorithms for allocating data in distributed database systems. Distrib. Parallel Databases **11**, 5–32 (2002)
3. Akal, F., Böhm, K., Schek, H.-J.: OLAP query evaluation in a database cluster: a performance study on intra-query parallelism. In: Manolopoulos, Y., Návrat, P. (eds.) ADBIS 2002. LNCS, vol. 2435, pp. 218–231. Springer, Heidelberg (2002)

4. Apers, P.M.G.: Data allocation in distributed database systems. ACM Trans. Database Syst. **13**(3), 263–304 (1988)
5. Bellatreche, L., Benkrid, S.: A joint design approach of partitioning and allocation in parallel data warehouses. In: Pedersen, T.B., Mohania, M.K., Tjoa, A.M. (eds.) DaWaK 2009. LNCS, vol. 5691, pp. 99–110. Springer, Heidelberg (2009)
6. Bellatreche, L., Benkrid, S., Crolotte, A., Cuzzocrea, A., Ghazal, A.: The F&A methodology and its experimental validation on a real-life parallel processing database system. In: CISIS'12, pp. 114–121 (2012)
7. Bellatreche, L., Boukhalfa, K.: An evolutionary approach to schema partitioning selection in a data warehouse. In: Tjoa, A.M., Trujillo, J. (eds.) DaWaK 2005. LNCS, vol. 3589, pp. 115–125. Springer, Heidelberg (2005)
8. Bellatreche, L., Boukhalfa, K., Richard, P.: Data partitioning in data warehouses: hardness study, heuristics and ORACLE validation. In: Song, I.-Y., Eder, J., Nguyen, T.M. (eds.) DaWaK 2008. LNCS, vol. 5182, pp. 87–96. Springer, Heidelberg (2008)
9. Bellatreche, L., Boukhalfa, K., Richard, P.: Referential horizontal partitioning selection problem in data warehouses: hardness study and selection algorithms. Int. J. Data Warehous. Min. **5**(4), 1–23 (2009)
10. Bellatreche, L., Cuzzocrea, A., Benkrid, S.: F&A: a methodology for effectively and efficiently designing parallel relational data warehouses on heterogenous database clusters. In: Pedersen, T.B., Mohania, M.K., Tjoa, A.M. (eds.) DaWak 2010. LNCS, vol. 6263, pp. 89–104. Springer, Heidelberg (2010)
11. Bellatreche, L., Cuzzocrea, A., Benkrid, S.: Effectively and efficiently designing and querying parallel relational data warehouses on heterogeneous database clusters: the F&A approach. J. Database Manage. **23**, 17–51 (2012)
12. Bergsten, B., Couprie, M., Valduriez, P.: Overview of parallel architectures for databases. Comput. J. **36**(8), 734–740 (1993)
13. Bezdek, J.C., Ehrlich, R., Full, W.: Fcm: the fuzzy c-means clustering algorithm. Comput. Geo-sci. **10**(2–3), 191–203 (1984)
14. Borr, A.: Transaction monitoring in encompass: reliable distributed transaction processing. In: Proceedings of the Very Large Database Conference, pp. 155–165. IEEE Press (1981)
15. Burkhard, W.A., Menon, J.: Disk array storage system reliability. In: FTCS, pp. 432–441 (1993)
16. Ceri, S., Negri, M., Pelagatti, G.: Horizontal data partitioning in database design. In: 1982 ACM SIGMOD International Conference on Management of Data, pp. 128–136 (1982)
17. Chang, R.-S., Chang, H.-P., Wang, Y.-T.: A dynamic weighted data replication strategy in data grids. In: Proceedings of the 2008 IEEE/ACS International Conference on Computer Systems and Applications, AICCSA '08, pp. 414–421. IEEE Computer Society, Washington, DC (2008)
18. Ciciani, B., Dias, D.M., Yu, P.S.: Analysis of replication in distributed database systems. IEEE Trans. Knowl. Data Eng. **2**, 247–261 (1990)
19. Copeland, G.P., Alexander, W., Boughter, E., Keller, T.: Data placement in bubba. In: ACM SIGMOD International Conference on Management of Data, pp. 99–108 (1988)
20. Costa, J.P., Furtado, P.: Poster session: towards a QoS-aware DBMS. In: ICDE Workshops, pp. 50–55 (2008)
21. Cuzzocrea, A.: Providing probabilistically-bounded approximate answers to non-holistic aggregate range queries in OLAP. In: 8th ACM International Workshop on Data Warehousing and OLAP (DOLAP 05), pp. 97–106 (2005)

22. Cuzzocrea, A.: Theoretical and practical aspects of warehousing, querying and mining sensor and streaming data. J. Comput. Syst. Sci. **79**(3), 309–311 (2013)
23. Cuzzocrea, A., Darmont, J., Mahboubi, H.: Fragmenting very large XML data warehouses via k-means clustering algorithm. Int. J. Bus. Intell. Data Min. **4**(3–4), 301–328 (2009)
24. Cuzzocrea, A., Mansmann, S.: OLAP visualization: models, issues, and techniques. In: Wang, J. (ed.) Encyclopedia of Data Warehousing and Mining, pp. 1439–1446. IGI Global, Hershey (2009)
25. Cuzzocrea, A., Russo, V., Saccà, D.: A robust sampling-based framework for privacy preserving OLAP. In: Song, I.-Y., Eder, J., Nguyen, T.M. (eds.) DaWaK 2008. LNCS, vol. 5182, pp. 97–114. Springer, Heidelberg (2008)
26. Cuzzocrea, A., Serafino, P.: LCS-Hist: taming massive high-dimensional data cube compression. In: 12th International Conference on Extending Database Technology (EDBT 09), pp. 768–779 (2009)
27. Cuzzocrea, A., Wang, W.: Approximate range-sum query answering on data cubes with probabilistic guarantees. J. Intell. Inf. Syst. **28**(2), 161–197 (2007)
28. Darabant, A.S., Campan, A.: Semi-supervised learning techniques: k-means clustering in OODB fragmentation. In: Second IEEE International Conference on Computational Cybernetics (ICCC 04), Vienna, Austria, pp. 333–338. IEEE Computer Society (2004)
29. Dewitt, D., Gerber, R.H., Graefe, G., Heytens, M.L., Kumar, K.B., Muralikrishna, M.: Gamma - a high performance dataflow database machine. VLDB **10**, 228–237 (1986)
30. DeWitt, D., Gray, J.: Parallel database systems: the future of high performance database systems. Commun. ACM **35**(6), 85–98 (1992)
31. DeWitt, D., Madden, S., Stonebraker, M.: How to build a high-performance data warehouse. http://db.lcs.mit.edu/madden/high_perf.pdf
32. Forestiero, A., Mastroianni, C., Spezzano, G.: Qos-based dissemination of content in grids. Future Gener. Comp. Syst. **24**(3), 235–244 (2008)
33. Furtado, P.: Experimental evidence on partitioning in parallel data warehouses. In: 7th ACM International Workshop on Data Warehousing and OLAP (DOLAP), pp. 23–30 (2004)
34. Furtado, P.: Efficient, chunk-replicated node partitioned data warehouses. In: ISPA, pp. 578–583 (2008)
35. Furtado, P.: Efficient and robust node-partitioned data warehouses. In: Erickson, J. (ed.) Database Technologies: Concepts, Methodologies, Tools, and Applications, pp. 658–677. IGI Global, IGI Global (2009)
36. Gorla, N., Yan, B.P.W.: Vertical fragmentation in databases using data-mining technique. In: Erickson, J. (ed.) Database Technologies: Concepts, Methodologies, Tools, and Applications, pp. 2543–2563. IGI Global, Hershey (2009)
37. Hababeh, I.O., Ramachandran, M., Bowring, N.: A high-performance computing method for data allocation in distributed database systems. J. Supercomput. **39**(1), 3–18 (2007)
38. Hsiao, H.-I., DeWitt, D.J.: Replicated data management in the gamma database machine. In: Workshop on the Management of Replicated Data, pp. 79–84 (1990)
39. Hsiao, H.-I., Dewitt, D.J.: Chained declustering: a new availability strategy for multiprocssor database machines. In: ICDE'90, pp. 456–465 (1990)
40. Coffman Jr., E.G., Leung, J.Y., Ting, D.W.: Bin packing: maximizing the number of pieces packed. Acta Inform. **9**, 263–271 (1978)

41. Karimi Adl, R., Rouhani Rankoohi, S.M.T.: A new ant colony optimization based algorithm for data allocation problem in distributed databases. Knowl. Inf. Syst. **20**(3), 349–373 (2009)

42. Lima, A.A.B., Mattoso, M., Valduriez, P.: Adaptive virtual partitioning for OLAP query processing in a database cluster. In: Lifschitz, S. (ed.) SBBD'04, Brasilia, Brésil, pp. 92–105 (2004)

43. Lima, A.B., Furtado, C., Valduriez, P., Mattoso, M.: Parallel olap query processing in database clusters with data replication. Distrib. Parallel Database J. **25**(1–2), 97–123 (2009)

44. Loukopoulos, T., Ahmad, I.: Static and adaptive distributed data replication using genetic algorithms. J. Parallel Distrib. Comput. **64**(11), 1270–1285 (2004)

45. Mansouri, Y., Monsefi, R.: Optimal number of replicas with qos assurance in data grid environment. In: Proceedings of the 2008 Second Asia International Conference on Modelling & Simulation (AMS), AMS '08, pp. 168–173. IEEE Computer Society, Washington, DC (2008)

46. Märtens, H., Rahm, E., Stöhr, T.: Dynamic query scheduling in parallel data warehouses: concurrency computation practice and experience. In: Monien, B., Feldmann, R.L. (eds.) Euro-Par 2002. LNCS, vol. 2400, pp. 321–331. Springer, Heidelberg (2002)

47. Menon, S.: Allocating fragments in distributed databases. IEEE Trans. Parallel Distrib. Syst. **16**(7), 577–585 (2005)

48. Nehme, R.V., Bruno, N.: Automated partitioning design in parallel database systems. In: ACM SIGMOD'11, pp. 1137–1148 (2011)

49. Noaman, A.Y., Barker, K.: A horizontal fragmentation algorithm for the fact relation in a distributed data warehouse. In: 8th International Conference on Information and Knowledge Management (CIKM'99), November 1999, pp. 154–161 (1999)

50. O'Neil, P., O'Neil, E.B., Chen, X.: The star schema benchmark (2007). http://www.cs.umb.edu/poneil/starschemab.pdf

51. Özsu, M.T., Valduriez, P.: Principles of Distributed Database Systems, 2nd edn. Prentice Hall, Englewood Cliffs (1999)

52. Page, T.H.: Tpc benchmark^{TM}d (decision support). http://www.tpc.org

53. Pavlo, A., Curino, C., Zdonik, S.: Skew-aware automatic database partitioning in shared-nothing, parallel OLTP systems. In: ACM SIGMOD'12, pp. 61–72. ACM, New York (2012)

54. Phan, T., Li, W.-S.: Load distribution of analytical query workloads for database cluster architectures. In: EDBT, pp. 169–180 (2008)

55. Rao, J., Zhang, C., Lohman, G., Megiddo, N.: Automating physical database design in a parallel database. In: ACM SIGMOD'02, June 2002, pp. 558–569 (2002)

56. Saccà, D., Wiederhold, G.: Database partitioning in a cluster of processors. ACM Trans. Database Syst. **10**(1), 29–56 (1985)

57. Sarathy, R., Shetty, B., Sen, A.: A constrained nonlinear 0–1 program for data allocation. Eur. J. Oper. Res. **102**(3), 626–647 (1997)

58. Stöhr, T., Märtens, H., Rahm, E.: Multi-dimensional database allocation for parallel data warehouses. In: VLDB'00, pp. 273–284 (2000)

59. Taniar, D., Leung, C.H.C., Rahayu, W., Goel, S.: High Performance Parallel Database Processing and Grid Databases. Wiley Publishing, Hoboken (2008)

60. Teradata. Dbc/1012 database computer system manual release 2.0. Technical document C10-0001-02 (1985)

61. Thiele, M., Bader, A., Lehner, W.: Multi-objective scheduling for real-time data warehouses. Comput. Sci. - R&D **24**(3), 137–151 (2009)

62. Wolfson, O., Milo, A.: The multicast policy and its relationship to replicated data placement. ACM Trans. Database Syst. **16**(1), 181–205 (1991)
63. Zhu, H., Gu, P., Wang, J.: Shifted declustering: a placement-ideal layout scheme for multi-way replication storage architecture. In: Proceedings of the 22nd Annual International Conference on Supercomputing, ICS '08, pp. 134–144. ACM, New York (2008)
64. Zilio, D.C., Jhingran, A., Padmanabhan, S.: Partitioning key selection for a shared-nothing parallel database system. In: IBM Research Report RC (1994)
65. Zilio, D.C., Rao, J., Lightstone, S., Lohman, G.M., Storm, A., Garcia-Arellano, C., Fadden, S.: DB2 design advisor: integrated automatic physical database design. In: Proceedings of the International Conference on Very Large Databases (VLDB), August 2004, pp. 1087–1097 (2004)

Improving Clustering-Based Schema Matching Using Latent Semantic Indexing

Alsayed Algergawy[1,2(✉)], Seham Moawed[3], Amany Sarhan[2],
Ali Eldosouky[3], and Gunter Saake[4]

[1] Institute of Computer Science, Friedrich Schiller University of Jena,
Jena, Germany
alsayed.algergawy@uni-jena.de
[2] Department of Computer Engineering, Tanta University, Tanta, Egypt
[3] Department of Computer Engineering, Mansoura University, Mansoura, Egypt
[4] Department of Computer Science, University of Magdeburg, Magdeburg, Germany

Abstract. The increasing size and the widespread use of XML data and different types of ontologies result in the big challenge of how to integrate these data. A critical step towards building this integration is to identify and discover semantically corresponding elements across heterogeneous data sets. This identification process becomes more and more challenging when dealing with large schemas and ontologies. Clustering-based matching is a great step towards more significant reduction of the search space and thus improving the matching efficiency. However, current methods used to identify similar clusters depend on literally matching terms. To keep high matching quality along with high matching efficiency, hidden semantic relationships among clusters' elements should be discovered. To this end, in this paper, we propose a Latent Semantic Indexing-based approach that allows retrieving the conceptual meaning between clusters. The experimental evaluations reveal that the proposed approach permits encouraging and significant improvements towards building large-scale matching approaches.

Keywords: Schema matching · Large-scale matching · Latent semantic indexing · Partitioning-based matching · Hierarchical clustering method · Vector Space Model (VSM) · Document similarity

1 Introduction

Schema matching is the task of identifying and discovering correspondences between semantically similar elements of two schemas or ontologies [31,33]. The demand for schema matching is high in a diverse number of data application scenarios, such as data integration [10,16] and web service discovery [4,20]. Due to heterogeneities inherent in schemas, manual matching becomes expensive, extremely tedious, and error prone. Therefore, efforts are invested in the development of automated schema matching systems. Furthermore, the rapidly increasing size and use of XML schemas and ontologies adds additional dimensions of challenges to cope with the large matching problem [30].

© Springer-Verlag Berlin Heidelberg 2014
A. Hameurlain et al. (Eds.): TLDKS XV, LNCS 8920, pp. 102–123, 2014.
DOI: 10.1007/978-3-662-45761-0_4

To deal with these challenges, several approaches have been designed to improve the performance of the matching process for large-scale schemas involving both matching aspects: *effectiveness* and *efficiency* [5,14,19,21,32,35]. These solutions include matching techniques that depend on the partition-based principle [2,14,21]. These partition-based matching techniques divide input schemas/ ontologies into a set of partitions and execute a partition-wise matching between the two schemas. The partitioning is performed in such a way that each partition of the first schema is matched with only a small subset of the partitions of the second schema (ideally, only with one partition) [30]. The entities of the dissimilar partition pairs can be eliminated from further matching process thus reducing the search space to achieve better efficiency. Space complexity of the matching process is also reduced. Reducing the search space of the matching process indeed achieves better matching efficiency, however, it does not guarantee the matching quality. Determining and selecting similar clusters for further matching plays an important role to keep high matching quality along with high matching efficiency.

To partition input schemas/ontologies, COMA++ uses relatively simple heuristic rules to partition the input schemas, often resulting in too few or too many partitions [14]. Both MOM and Falcon have been applied only to certain ontology languages and cannot be applied to other data models [21,35]. Algergawy et al. use a bottom-up clustering scheme which utilizes the context-based structural node similarities [2]. To determine similar partitions, COMA++ only uses limited information about the partition (only the root node of the partition) to determine the similarity between partitions of the input schemas. On the other hand, solutions, such as Falcon [21], fully evaluate the input ontologies to assess the partition similarity. In Algergawy et al. [2], a light-weight similarity measure is applied that considers all elements of each cluster pair and represents each cluster as a cluster document. It uses the Vector Space Model and TF-IDF to determine the similarity between cluster documents.

Unfortunately, the Vector Space Model (VSM) depends upon literally matching document terms with those appearing in a query [8]. The inaccuracy of lexical matching methods is coming from the inability to determine concepts between documents and the query. So, the literal terms in a user's query may not match those of a relevant document (synonymy). In addition, most words have multiple meanings (polysemy), so terms in a user's query will literally match terms in irrelevant documents. Latent semantic indexing (LSI) is a more suitable approach that allows retrieving information on the basis of a conceptual topic or meaning of a document [12,22]. To this end, in this paper, we capture features introduced by the latent semantic indexing technique in large-scale schema matching problems. In particular, we first represent input schemas as rooted labelled trees, called *schema trees*. The use of a common data structure, *schema tree*, to model input schemas, enables matching among different schemas and ontologies. We then develop an agglomerative clustering algorithm to partition each schema tree into a set of disjoint groups. The clustering algorithms depends on the structural properties of the schema tree. To identify and determine similar clusters across two cluster sets representing two schema trees, we develop an LSI-based technique which is able to discover hidden semantic relationships between similar clusters. Once having similar clusters, we finally apply a set of element matchers to get correspondences between their elements. To verify the performance of the proposed approach, we conducted a set of experiments in order to prove its superiority upon previous work.

To sum up, the main contributions of the paper can be stated as follows:

- addressing the problem of partitioning-based schema matching,
- developing and elaborating an XSOM-based parser to facilitate XML data representation,
- proposing an LSI-based approach to determine similar clusters in the context of schema matching, and
- conducting an intensive set of experiments to validate the proposed approach.

The rest of the paper is structured as follows. Related work is presented in Sect. 2. We describe latent semantic indexing in Sect. 3. We then introduce the proposed matching framework in Sect. 4, concentrating on similar clusters identification. We report experiments conducted and analysis results in Sect. 5. Section 6 concludes the paper.

2 Related Work

Semantic heterogeneity is a key problem in different data sharing systems, be it a federated database [6], a data integration system [15,16], a web service [20], or a peer data management system [18]. Involved data sources are typically designed independently, and hence use different schemas. To obtain meaningful interoperation, one needs a semantic mapping between the schemas, i.e. a set of expressions that specify how the data in one source corresponds to the data in the other. Hence, the specific problem of schema matching has to be addressed before mapping is constructed. To this aim, a set of correspondences among similar elements in different schemas has to be identified. Manually constructing a match is a very labor intensive task that requires complete knowledge of the semantics of the data in the schemas being matched. Solutions that try to provide some automatic support for schema matching have received steady attention over the years [7,31,33].

Unfortunately, most of these systems severely lack performance when dealing with large matching problems. Consequently, several approaches have been proposed to address the problem of matching two large schemas [2,14,19,21, 30,32,35]. Promising areas for large-scale schema matching lie in four main directions: reduction of search space for matching, parallel matching, self-tuning match workflows and reuse of previous match results [30]. In this section, we pay great attention to the approaches that perform reduction of the search space. The standard approach of cross join evaluation for schema matching reduces match efficiency and quality. In order to reduce the search space for matching, two methods can be used: early pruning of dissimilar element pairs and partition-based matching.

Quick ontology matching (QOM) was one of the first approaches to implement the idea of early pruning of dissimilar element pairs [17]. It iteratively applies a sequence of matchers and can restrict the search space for every matcher. Peukert et al. introduce a set of filter operators within match workflows to prune dissimilar element pairs (whose similarity is below some minimal threshold) from intermediate match results [28]. They also propose a rule-based approach to rewrite match workflows for improving efficiency, in particular by placing filter operators within sequences of matchers [27].

COMA++ was one of the first systems to support partition-based schema matching [14]. It depends on fragment matching which has two phases. The first phase determines fragments of the two schemas and identifies the most similar ones. Detecting similar fragments is some kind of light-weight matching via the similarity of fragment roots. The second phase identifies corresponding elements between each pair of similar fragments. Finally, the fragment-based match results are merged to obtain the complete output mapping [14].

Another matching system that supports partition-based matching is Falcon-AO [21]. It initially partitions the ontologies into relatively small disjoint blocks by using structural clustering. Then, matching is applied to the most similar blocks from the two ontologies. To determine block similarity, the algorithm utilizes the so-called anchors. Anchors are highly similar element pairs that are determined before partitioning by a combined name/comment matcher. Dynamic partition-based matching is supported by AnchorFlood [32]. It avoids the a-priori partitioning of the ontologies by utilizing anchors (similar concept pairs). It takes them as a starting point to incrementally match elements in their structural neighborhood until no further matches are found or all elements are processed. Thus the partitions (segments) are located around the anchors.

Zhong et al. propose an unbalanced ontology matching approach, which concerns matching a lightweight ontology with a more heavyweight one [36]. They abstract the subontology (partition) from the heavyweight ontology that is most similar to the smaller one and consider this sub-ontology for matching. To determine this sub-ontology, the approach needs to carry out a nested loop to determine the similarity values between concepts from the two ontologies. To this end, name-based similarity measures such as Edit distance and WordNet have been used. Concepts from the larger ontology with similarity values higher than a predefined threshold are then selected. Finally, the subontology is determined by evaluating the subgraphs around the similar elements found in the first step. We observe that in order to determine a similar sub-ontology, whole concepts from two ontologies have to be compared using name-based similarity measures, which is not efficient for large matching problems.

Algergawy et al. uses a clustering-based matching approach that is based on an agglomerative bottom-up hierarchical fashion [2]. It is generic and can be applied to different data models including XML schemas. The clustering scheme is performed based on the context-based structural node similarities. Then, a light weight linguistic technique is used to find similar partitions to match.

This technique makes use of the Vector Space Model (VSM) for computing the similarity between clusters.

To sum up, partitioning-based matching techniques improve the matching efficiency, however, they do not guarantee a high matching quality. Identifying and selecting similar partitions for matching plays an important role in this aspect. To the best of our knowledge, most of current matching techniques ignore this role. Therefore and in order to address these challenges, we introduce a new LSI-based approach to correctly identify and select the similar clusters.

3   Latent Semantic Indexing

One typical scenario of human machine interaction in information retrieval is by natural language queries: the user formulates a request, e.g., by providing a number of keywords or some free-form text, and expects the system to return the relevant data in some amenable representation, e.g., in form of a ranked list of relevant documents. Many retrieval methods are based on simple word matching strategies to determine the rank of relevance of a document with respect to a query. It is well known that literal term matching has severe drawbacks, mainly due to the ambivalence of words and their unavoidable lack of precision as well as due to personal style and individual differences in word usage.

A popular approach that depends on literal term matching is the Vector Space Model (VSM) [8,12]. The vector space model procedure can be divided into three stages. The first stage is the document indexing where content bearing terms are extracted from the document text. The second stage is the weighting of the indexed terms to enhance retrieval of document relevant to the user. The last stage ranks the document with respect to the query according to a similarity measure. The VSM considers the terms in documents as being independent from each other, an assumption which is never satisfied by the human language. An idea can be expressed in many ways (synonymy) and, moreover, many words may have multiple meanings (polysemy).

Latent Semantic Indexing (LSI) [12,22] is a statistical technique which tries to surpass some limitations imposed by the traditional Vector Space Model (VSM). It exploits the dependencies between words by assuming that there is some underlying or "latent" structure in word usage across documents that is partially obscured by variability in word choice and this structure can be revealed statistically.

LSI projects queries and documents into a space with "latent" semantic dimensions. In the latent semantic space, a query and a document can have high cosine similarity even if they do not share any terms. We can look at LSI as a similarity metric that is an alternative to word overlap measures like *tf.idf* [25]. LSI usually takes the (high dimensional) vector space representation of documents based on term frequencies [14] as a starting point and applies a dimension reducing linear projection. The specific form of this mapping is determined by a given document collection and is based on a Singular Value Decomposition (SVD) of the corresponding term/document matrix. The general claim is that similarities between documents or between documents and queries can be more reliably estimated in the reduced latent space representation than in the original representation. The rationale is that documents which share frequently co-occurring terms will have a similar representation in the latent space, even if they have no terms in common. LSI thus performs some sort of noise reduction and has the potential benefit to detect synonyms as well as words that refer to the same topic. In many applications this has proven to result in more robust word processing.

To make the paper self-contained, in the following, we present main steps of LSI [22]:

- **Constructing Term Document Matrix.** Each term is represented by a row and each document is represented by a column. Initially, each cell a_{ij} in the matrix A is represented by the number of times the associated term appears in the indicated document, tf_{ij}. Once the matrix is created, local and global weighting functions can be applied to each non-zero element in the matrix. The weighting functions transform each cell, a_{ij} of A, to be the product of a local term weight which describes the relative frequency of a term in a document, and a global weight, g_i, which describes the relative frequency of the term within the entire collection of documents. The local weighting function of $\log(tf_{if} + 1)$ decreases the effect of large differences in frequencies. The global weighting function of Entropy, which is defined as $1 + \sum_j \frac{P_{ij} \log(P_{ij})}{\log(n)}$ where $P_{ij} = \frac{tf_{ij}}{gf_i}$, is the total number of times the term appears in the entire collection of n documents, gives less weight to terms occurring frequently in a document collection. Therefore, each non-zero element in the term-document matrix is represented as:

$$a_{ij} = (1 + \sum_j \frac{P_{ij} \log(P_{ij})}{\log(n)}) \times \log(tf_{ij} + 1). \tag{1}$$

- **Decomposing the Term Document Matrix.** LSI applies singular value decomposition (SVD) to the matrix A. In SVD, a rectangular matrix is factored into the product of other three matrices as in

$$A = USV^T \tag{2}$$

where U is an $m \times m$ orthogonal matrix, $U^T U = I_m$, V is an $n \times n$ orthogonal matrix, $V^T V = I_n$, and S is a diagonal matrix of decreasing singular values such that $s_{1,1} \geq s_{2,2}... \geq s_{r,r} > 0$, and $s_{i,j} = 0$ where $i \neq j$. I_m and I_n are the identity matrices of orders m and n, respectively. The matrix U gives a vector for each term in LSI space, while the matrix V represents each document as a vector.

- **Dimensionality Reduction.** In LSI, it is not the intent to reproduce A. The main goal is to retain the largest singular values. In the literature, this is called dimensionality reduction. LSI computes a low rank approximation to A using a truncated SVD [22]. Let k be an integer and $k \ll min(m, n)$, U_k is defined to be the first k columns of U, and V_k^T to be the first k rows of V^T. Let $S_k = diag[s_1, ..., s_k]$ contain the first k largest singular values as in the following equation:

$$A_k = U_k S_k V_k^T \qquad (3)$$

This is a new pseudo term-document matrix with reduced dimension. The SVD operation, along with this reduction, has the effect of preserving the most important semantic information in the text while reducing noise and other undesirable artifacts of the original space of A.

– **Incorporating the Query and Ranking the Documents.** A query, similar to a document, is a set of words which must be represented as a vector in the k-dimensional space. It can be represented as:

$$q = q^T U_k S_k^{-1} \qquad (4)$$

where q is the vector of words in the users query, multiplied by the appropriate term weights. The sum of these $k-$dimensional terms vectors is reflected by the term $= q^T U_k$ and the right multiplication by S_k^{-1} differentially weights the separate dimensions. Thus, the query vector is located at the weighted sum of its constituent term vectors. The query vector can then be compared to all existing document vectors, and the documents ranked by their similarity to the query. A common similarity measure can be used to reflect the relationship between the query vector and every document vector. Typically, the results are ranked and top-k documents *or* documents exceeding some cosine threshold are returned to the user.

4   The Matching Framework

In this section, we introduce the proposed schema matching framework. The framework consists of four main steps, as shown in Fig. 1. In the following, we describe each step focusing on the parsing and similar cluster determination steps.

4.1   Data Model and Schema Preparation

XML is a flexible modeling language with self-explanatory tags that allow the storage of information in semi-structured formats [1]. There are two types of XML data: XML schema and XML document. An XML schema allows describing the structure and the legal building blocks for an XML document, while an XML document (document instance) represents a snapshot of what the XML document contains. Several XML schema languages have been proposed [23]. Among them, XML document type definition (DTD) and XML Schema Definition (XSD) are commonly used. DTD has limited capabilities compared to other schema languages, such as XSD. Moreover, XML schema definition (XSD) aims to be more expressive than DTD and more usable by a wider variety of

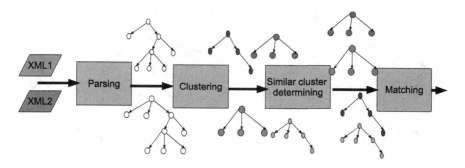

Fig. 1. Schema matching steps.

applications such as XQuery[1], SOAP, and web services[2]. Therefore, through the paper, we use the term "schema" to denote XML schema (XSD).

An XML schema consists of a set of components. The XML schema components can be broadly classified into three main groups as described below:

- **Primary components** may or must have names and include the following components: simple type definitions, complex type definitions, element declarations, and attribute declarations. The element and attribute declarations must have names, while the type definitions may have names.
- **Secondary components** must have names. Attribute group definitions, identify constraint definitions, model group definitions, notation declarations, type alternatives, and assertions are examples of such components.
- **Helper components** provide small parts of other components, they are not independent of their context and contain components such as annotations, model groups, particles, wildcards, and attribute use.

To make the proposed approach more generic, the input XML schemas should be internally represented using a common data model. The choice of which data model should be used is an important step towards building a reasonable schema matching system. The data model should be able to normalize schemas that are represented by different schema languages, thus eliminating syntax differences between schemas. Most current schema matching systems choose graph data structure as the internal representation [31,33]. The choice of graphs as an internal representation for the schemas to be matched has many motivations. First, graphs are well-known data structures and have their algorithms and implementations. Second, by using the graph as a common data model, the schema matching problem is transformed into another standard problem; graph matching. XML schemas can also be represented as trees by dealing with nesting and repetition problems using a set of predefined transformation rules [24].

In our implementation, we represent XML schemas as rooted, labeled trees, called *schema trees, ST* [5]. A schema tree consists of a finite set of nodes and

---

[1] http://www.w3.org/TR/xquery/.

[2] http://msdn.microsoft.com/en-us/library/ee265410(v=bts.10).aspx.

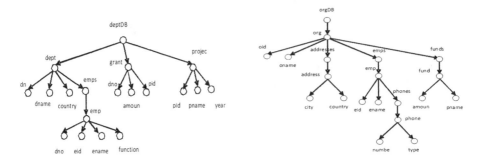

Fig. 2. Schema tree, *deptDB*.   **Fig. 3.** Schema tree, *orgDB*.

a finite set of edges. Each node is uniquely identified by an object identifier and expresses the component' features, such as element name and element datatype, while each edge represents the relationship between every two nodes. Figures 2 and 3 present the schema tree representation of two XML schemas taken from [9]. Both *deptDB* and *orgDB* represent information about departments with their employees and grants, as well as the projects for which grants are awarded. The figures show the tree representation of the two schemas, wherein each node is associated with the name label, such as *grant* and *funds* from *deptDB* and *orgDB*, respectively.

We use the XML Schema Object Model (XSOM) parser[3] to parse input XML schemas. XSOM is a Java library that allows applications to easily parse XML schema documents and to inspect information in them. The library is a simple and effective implementation of "schema components" as defined in the XML schema. The parsing process starts by defining a new class using the XSOM. Through the constructor, we create an empty tree which will be filled with schema elements extracted through the parsing operation. As in [3,5], we classify schema tree nodes into two kinds: *atomic nodes* and *complex nodes*. Atomic nodes are the leaf nodes in the schema tree while complex nodes are the internal nodes inside the tree. We then instantiate an object of the XSOM parser through a defined class. This class enables us to get all constructs of the schema and related schemas and put them into memory for further processing, by using the defined object methods. Once the schema component is resolved, we iterate through global declarations inside the root element to iteratively build the corresponding schema tree.

4.2   Schema Clustering

Once a schema is parsed and internally represented as a schema tree, the next step is to divide it into a set of disjoint sub-trees. By this step, we aim to simplify the

---

[3] https://xsom.java.net.

matching processing, especially when dealing with large-scale schemas. To this end, we make use of our clustering algorithm presented in [2]. To make the paper self-contained, we briefly present the algorithm. Clustering is a useful technique for grouping nodes such that nodes within a single cluster are structurally similar, while nodes in different groups are dissimilar. First, we introduce the *node context*, which is defined as the node surroundings. This means that the context of a node, $C(v_i)$, is the combination of the node itself as well as all parents and children of the node. Based on the node context, we then compute the structure similarity between every pair of nodes in the schema tree.

The structure similarity between two nodes v_i and v_j which exist in the same ST is computed based on the number of common nodes between their contexts, $|C(v_i) \cap C(v_j)|$. Based on this structural similarity, we construct a link between each node pair, containing the two nodes and their structural similarity. The set of generated links constitutes a hash table called the *links hash table*. By using a threshold value greater than 0 we can dramatically reduce the number of entries in the links hash table. It should be noted that the similarity is assumed to be 0 if there is no pre-computed link. This table is used as an input for the clustering algorithm.

We develop an agglomerative clustering algorithm, which produces a tree representing the hierarchy of clusters in a bottom-up way. The algorithm mainly consists of the following four steps:

1. *Preparation.* The structural similarity is computed and the links hash table is then constructed.
2. *Cluster initialization.* In this step, the bottom level of the cluster hierarchy is developed by representing each node as a cluster.
3. *Cluster hierarchy construction.* This is the main step of the clustering algorithm. It is devoted to build the cluster hierarchy by merging elements from different clusters to form one cluster based on specified merging criteria.
4. *Best cluster selection.* It selects the cluster solution. More information can be found in [2].

Example 1. By applying the clustering algorithm to the two schema trees illustrated in Figs. 2 and 3, we get two cluster sets. $CSet_1 = \{C_{11}, C_{12}\}$ and $CSet_2 = \{C_{21}\}$ for *deptDB* and *orgDB* schemas, respectively, as shown in Fig. 4. The figure indicates that *deptDB* is partitioned into two semantically structured clusters. The first, C_{11}, represents projects and their funds, while the second, C_{12}, represents departments and employees working on these projects. Figure 4 also shows that the *orgDB* schema is not partitioned since the structural organization of the schema is not semantically clear like the *deptDB* schema. This example shows the ability of the clustering algorithm to correctly cluster schema trees into semantically structured partitions.

4.3   Similar Cluster Determination

The proposed approach focuses on 2-way or pairwise schema matching where two related input schemas are matched with each other. As mentioned before, the

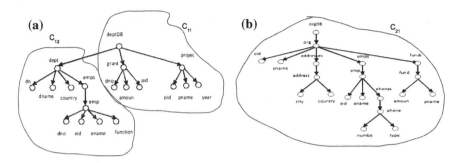

Fig. 4. Schema tree partitions

clustering algorithm divides each schema tree into a set of clusters. Each cluster contains a set of nodes that are structurally similar. The task is to determine which sets of clusters are similar. This information is used later as input for the matching algorithm.

Latent semantic indexing aims to detect semantically similar partitions (clusters) in the two schema trees. The motivation here is to reduce the match overhead by applying matching on similar partitions only and ignoring the irrelevant ones. Algorithm 1 is proposed to achieve this task. It accepts two sets of clusters as input and processes them to determine similar clusters across the two sets. The algorithm has the following main steps, as shown in Algorithm 1.

– *Preparation of term-document matrix.* The algorithm starts by initializing the similar cluster set, *Sim_Clust*, by setting it to the empty set, *line 1*. Then, to construct the document term matrix, all elements in the first cluster set ($CSet_1$) are extracted and analyzed. A set of normalization processes has been applied to the element names in order to obtain non-repeating terms in the cluster set. The normalization process has the following steps:
  • Tokenization. Each element name inside the cluster is parsed into a set of tokens using delimiters, such as punctuation, uppercase or special symbols, etc. For instance, *deptDB* → {*dept, DB*}.
  • Elimination. Tokens that are neither letters nor digits are eliminated.
  We have thus created the term vector by collecting the names of the nodes. As a next step, the term-document matrix, A, is initially created with each matrix cell representing the number of times the associated node name appears in the indicated cluster document, *line 2*. Finally, we apply the log-entropy weighting function, Eq. 1 on each entry in the document matrix, *line 3*. Thus, the term-document matrix is ready for the next stage.
  It should be noted that other normalization techniques are needed, such as the expansion method, especially when schema element names are too short. *dno* and *pid* are examples that are needed to be expanded.
– *Applying singular value decomposition.* To construct a semantic space wherein the names of the nodes in the $CSet_1$ and clusters that are closely associated are placed near one another, we apply the singular value decomposition technique.

Algorithm 1. Similar clustering determination

Require: Two sets of clusters, $CSet_1 = \{C_{11}, C_{12}, ..., C_{1n}\}$ and $CSet_2 = \{C_{21}, C_{22}, ..., C_{2m}\}$

Ensure: A set of similar clusters, $Sim\_Clust = \{(C_{1i}, C_{2j})|C_{1i} \in CSet_1, C_{2j} \in CSet_2\}$

{// **Step 1: Preparation**}
1: $Sim\_Clust \Leftarrow \emptyset$;
2: $A \Leftarrow analysis(CSet_1)$;
3: Compute for each entry in A :a_{ij}
   $a_{ij} \Leftarrow (1 + \sum_j \frac{P_{ij}\log(P_{ij})}{\log(n)}) \times \log(tf_{ij} + 1)$
   {// **Step 2: Singular Value Decomposition & reduction**}
4: Apply SVD to A : $A = USV^T$
5: Dimensionality reduction: $A_k = U_k S_k V_k^T$
   {// **Step 3: Query incorporating and folding**}
6: $Q \Leftarrow analysis(CSet_2)$;
7: $Q_k \Leftarrow Q^T U_k S_k^{-1}$;
   {// **Step 4: Similarity calculating and ranking**}
8: **for** $column_j \in Q_k$ **do**
9:    $q_j \Leftarrow Q_k(j)$;
10:   **for** $column_i \in A_k$ **do**
11:      $d_i \Leftarrow A_k(i)$;
12:      $simMat[i][j] = sim(q_j, d_i)$;
13:      **if** $simMat[i][j] \geq threshold$ **then**
14:         $Sim\_Clust.put(C_{1i}, C_{2j})$
15:      **end if**
16:   **end for**
17: **end for**

The technique factorizes a term-document matrix into its left singular vectors, right singular vectors, and singular values, *line 4*. Each node name within the cluster set is now represented by a singular vector via matrix U. Additionally, each cluster is represented by a singular vector via matrix V.

- *Reduction.* To reduce the noise and redundancy, LSI uses a truncated SVD, *line 5*, which consists in retaining only the largest k singular values and deleting the remaining ones which are smaller and thus considered unimportant. The columns corresponding to the small singular values are also removed from U and V. So, SVD allows the arrangement of the space to reflect the major associative patterns in the data, and ignore the smaller, less important influences. As a result, terms that do not actually appear in a document may still up close to the document, if that is consistent with the major patterns of association in the data.

- *Folding.* The following step is to prepare a set of clusters in the second cluster set $CSet_2$. Each cluster is treated as a user query. First, we analyze the element names of each cluster and we apply the same normalization process applied before on elements of the first cluster set to the second cluster set elements.

The query is then treated as an ordinary document and hence it should be put with new coordinates in the reduced $k - dimensional$ space, *lines 6&7*.

– *Calculating similarities*. Now, the two cluster sets have been prepared for comparison: one as a set of vectors via the matrix V, the second as a set of vector via the query matrix Q. Each vector in the two matrices represents a cluster. The current task is to compute the similarity between two sets of clusters and select similar clusters. As shown in the algorithm, *lines 8 to 17*, a query vector is extracted and compared with all the other cluster set elements. A cosine measure is used to compute the similarity between two vectors. If the computed similarity exceeds a specified threshold, the two clusters constitute a similar cluster pair to be then added to the final result, *Sim_Clust*.

The computed similarities between cluster pairs of the two schemas are used to construct a so-called cluster similarity matrix, *line 12*. If the computed similarity between each two clusters exceeds a specific threshold, the two clusters are put in the similar cluster set, *lines 13&14*.

4.4   Walk-Through Example

We provide an example that elaborates the proposed algorithms and gives more details about how to determine similar clusters. In this example, we use two schema trees illustrated in Figs. 2 and 3. We formulate the problem in this example as follows: given two cluster sets $CSet_1 = \{C_{11}, C_{12}\}$ and $CSet_2 = \{C_{21}\}$ shown in Fig. 4, identify similar clusters across the two cluster sets.

– Step 1: We select the cluster set of larger number of clusters, $CSet_1$, to construct the term-document matrix. After applying the normalization process on element names, we get the matrix A, as shown below. Each element in the matrix shows the number of occurrences of each term in the associated cluster (document). After getting the matrix A, we apply the log-entropy weighting scheme to get $A_{entropy}$.

$$
\begin{bmatrix}
dept \\
DB \\
dno \\
dname \\
country \\
emps \\
emp \\
eid \\
ename \\
function \\
grant \\
amount \\
pid \\
project \\
pname \\
year
\end{bmatrix}
A =
\begin{bmatrix}
1 & 1 \\
0 & 1 \\
2 & 1 \\
1 & 0 \\
1 & 0 \\
1 & 0 \\
1 & 0 \\
1 & 0 \\
1 & 0 \\
1 & 0 \\
0 & 1 \\
0 & 1 \\
0 & 2 \\
0 & 1 \\
0 & 1 \\
0 & 1
\end{bmatrix}
\Rightarrow A_{entropy} =
\begin{bmatrix}
0 & 0 \\
0 & 0.69 \\
0.09 & 0.06 \\
0.69 & 0 \\
0.69 & 0 \\
0.69 & 0 \\
0.69 & 0 \\
0.69 & 0 \\
0.69 & 0 \\
0.69 & 0 \\
0 & 0.69 \\
0 & 0.69 \\
0 & 1.1 \\
0 & 0.69 \\
0 & 0.69 \\
0 & 0.69
\end{bmatrix}
$$

- Step 2: The log-entropy matrix is then decomposed into three matrices as given by Eq. 2, where U is an 16×16 orthogonal matrix, S is an 16×2 diagonal matrix, and V is an 2×2 orthogonal matrix. Since S has only two eigenvalues, the SVD method should reserve only two columns in U and neglect the rest, and S should be limited only to two rows, and its dimensions should be truncated. After applying the SVD scheme, we get the following V and S matrices (U is not presented to save space since it is an 16×2 matrix).

$$V = \begin{bmatrix} -0.007 & -1 \\ -1 & 0.007 \end{bmatrix} S = \begin{bmatrix} 2.02 & 0 \\ 0 & 1.84 \end{bmatrix}$$

Low rank approximation to A, called A_k, can be created through the truncated SVD, via Eq. 3. For truncation, we assume to truncate 98 % of the singular values. In this example, there is no truncation and the matrices remain the same.

- Step 3: Incorporating the query, we incorporate the clusters of the *orgDB* schema into the new dimensional space created by SVD and its reduced form processes. The schema is partitioned into one cluster according to the applied threshold. Hence, we have one query which is analyzed and presented as an $m \times 1$ matrix. This matrix is projected onto the reduced term-document space via Eq. 4. The new coordinates of this query are represented in vector q, where $q = [-0.131 - 0.262]$.

- Step 4: The final step is applying cosine similarity function and ranking the documents as follows.

$sim(C_{11}, C_{21}) = sim(d_1, q) = 0.897$ and $sim(C_{12}, C_{21}) = sim(d_2, q) = 0.442$

Solving the same example using VSM yields the following results:

$sim_{VSM}(C_{11}, C_{21}) = 0.373$ and $sim_{VSM}(C_{12}, C_{21}) = 0.224$. If we set a threshold value of 0.3, then we get $Sim\_Clust(C_{11}, C_{21}), (C_{12}, C_{21})$ using the LSI-based method, while the similar cluster set contains only one similar cluster using the VSM-based method. From this example, it has been shown that the computed similarities by LSI are higher than those computed by VSM due to the ability of LSI to correlate semantically related terms that are latent in the collection of documents. Furthermore, the documents as well as the query vectors are represented by the new dimensions with semantic correlation between them.

4.5   Match Similar Clusters

Once settling on the similar clusters of the two schemas, the next step is to fully match similar clusters to obtain the correspondences between their elements. Each pair of the similar clusters represents an individual match task that is independently solved. Match results of these individual tasks are then combined to a single mapping, which represents the final match result. Since the matching part is not the main focus on the paper, we employ both name and type similarity measures to quantify the similarity between two similar cluster elements [3]. We simply introduce the two similarity measures (interested readers can refer to [3] for more details).

- *Name similarity measure:* Element names are considered important information sources for schema matching. Each element name should be normalized into a set of tokens and a set of string similarity measures can be applied on these tokens. Based on results presented in [3], we employ three string-based measures, namely, *Levenstein distance, N-gram distance,* and *Jaro similarity* [11].
- *Type similarity measure:* Although the element name is considered a necessary source for determining the element similarity, the consideration for other features also plays a different role. The element data type is another schema information that makes a contribution in determining the element similarity. The type similarity measure aids to prune some of the false positive matches produced from the name similarity measure. XML schema data types are divided into 12 communal types[4]. Therefore, in this paper, we build a data type similarity table. We calculate the similarity value for each data type pair based on the constraining facets of XML schema[5]. For more details, refer to [34].

Once the similarity between elements from two similar clusters has been computed using the name and type matchers, a weighted sum function is used to aggregate these similarity values. Elements with similarity values higher than a predefined threshold are selected as partial matching results. Finally, partial results from all similar clusters are combined to produce the final matching result.

5   Experimental Evaluation

To evaluate the effectiveness of the proposed approach, we conducted a set of experiments utilizing real-world schemas and ontologies. We ran all our experiments on 2.67 GHz Intel (R) Core i5 processor with 4 GB RAM running Windows 7. The proposed approach has been developed and implemented in Java.

5.1   Data Set

We collected data sets from different domains with different characteristics, as shown in Table 1. The table illustrates that the collected schemas are from 8 different domains[6] with different sizes ranging from small to large schemas. Within each domain, we use two schemas in order to apply the proposed approach. We choose these data sets to demonstrate the applicability of our approach to different data sources having different characteristics. More details about data sets in Table 1 can be found in [14,29].

---

[4] http://www.w3.org/TR/xmlschema-2/.

[5] XML Schema - Data Types Quick Reference, http://www.xml.dvint.com/.

[6] http://queens.db.toronto.edu/project/clio/index.php#testschemas.

Table 1. Data set specification.

Domain	Tested sources	No. of elements
Spicy	deptDB/orgDB	19/20
University	Uni1/Uni2	11/11
Web	Yahoo/ebay	37/37
TPC_ H	TPC_H1/TPC_H2	43/17
Finance	finan1/finan2	14/14
GeneX	GeneX1/GeneX2	75/85
Mondial	Mondial1/Mondail2	117/108
PO(large)	OpenTran_Invoice/OpenTran_Order	1113/1162

5.2 Evaluation Criteria

In our implementation, we consider two levels of evaluations, which can help answering the following two questions:

- Which is the better technique to determine similar clusters; LSI-based or VSM-based?
- What is the effect of both LSI-based and VSM-based techniques on the overall matching quality?

In order to answer these questions, we use the same criteria used in literature in terms of *precision, recall,* and *F-measure* [8]. In general, precision P can be defined as the degree of correctness of the result. In answering the first question, the result means the similar clusters, while in the second question it means match results (i.e., correspondences). It measures the ratio of correctly identified results (true positives, t_P) over the total number of identified results (true positives plus false positives f_P). It can be computed as: $P = \frac{t_P}{t_P+f_P}$. Recall, R, assesses the degree of completeness of the system. It measures the ratio of correctly identified results (true positives, t_P) over the total number of correct results (true positives plus false negatives f_N). It can be computed as: $R = \frac{t_P}{t_P+f_N}$.

However, neither precision nor recall alone can accurately assess the matching quality [13]. Precision evaluates the post-match effort that is needed to remove false positives, while recall evaluates the post-match effort that is needed to add true negatives to the final match result. Hence, it is necessary to consider a trade-off between them. There are several methods to handle such a trade-off, one of them is to combine both measures. The mostly used combined measure is *F-measure*. *F-measure* is the weighted harmonic mean of precision and recall. The traditional F-measure can be defined as:

$$F\text{-}measure = 2 \times \frac{P \times R}{P + R} \qquad (5)$$

5.3   Experimental Results

LSI-Based Approach Quality. We conducted two sets of experiments to validate the performance of the proposed approach and to answer the mentioned questions. The first set is devoted to answer the question "whether LSI-based or VSM-based technique is better in determining similar clusters in the context of partitioning-based schema matching". We validated the proposed approach using XML schemas illustrated in Table 1. Each XML schema is parsed and represented as a schema tree. The clustering-based approach, in [2], is applied to partition each schema tree into a set of clusters. To determine similar clusters among two sets of clusters, we applied both our LSI-based approach and the VSM-based approach [2]. The elements of cluster similarity matrix are ranked according to their similarity to each other and the similar clusters have been selected when their similarities exceed a predefined threshold. Results are summarized in Fig. 5.

Results represented in Fig. 5 can be classified into three main categories. The first one considering the University (Spicy and Finance which are not drawn) and TPC_H schemas, as shown in Fig. 5(a,b), illustrates that F-measure has its best values at low threshold and it decreases with increasing threshold values. The LSI-based method has an F-measure of nearly 1 over threshold values ranging between 0 and 0.5, and then the F-measure decreases to reach zero at threshold of 0.6 (for University) and 0.9 (for TPC). However, the VSM-based method has its best value at only two threshold values and it decreases to reach zero at a threshold value of 0.4 (for both schemas). This can be explained as both the

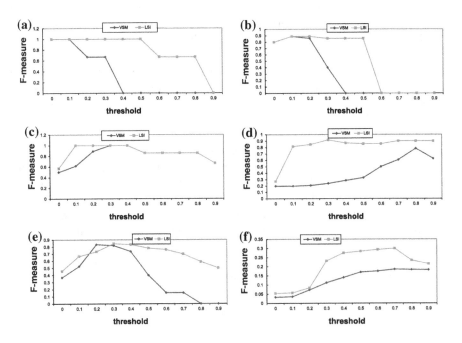

Fig. 5. Similar clusters quality

University (Spicy) and TPC_H have higher name heterogeneities, which make the VSM-based method fail to determine the correct similar clusters.

The second category considers the Web and Mondial schemas, as shown in Fig. 5(c,d). These schemas have lower name heterogeneities, which give the VSM-based method the chance to correctly identify similar clusters. The figures also show that as the threshold increases, the F-measure increases. However, the LSI-based method has higher F-measure values than the VSM-based method, especially for the Mondial schema.

The third category represents the PO schemas. It is the most difficult match task, since these schemas are highly heterogeneous. Therefore, the quality of similar cluster determination is lower compared to the other cases. However, the LSI-based method keeps higher F-measure values than the VSM-based method.

To sum up, Fig. 5 shows that the LSI-based technique outperforms the VSM-based technique across tested schemas. The figure also illustrates the capability of the LSI-based method to cover the hidden semantic relationships between cluster elements, which results in more accurate and quality similar clusters determination.

Furthermore, we compared the LSI-based and the VSM-based method with respect to the number of produced similar clusters. Based on results reported in Fig. 5, we found that the "best" similar clusters occur at different threshold values. Therefore, we decide to select a suitable threshold value to conduct this comparison. To this end, we select the similar clusters produced at a threshold value of 0.3. Results are reported in Table 2. The table represents the number of generated clusters after applying the clustering algorithm, the number of real similar clusters, the number of similar cluster (both correct and total numbers) generated by both techniques and its quality (F-measure). The table also verifies the results presented in Fig. 5. Table 2 illustrates that the LSI-based method outperforms the VSM-based method. This is can be explained due to the ability

Table 2. Comparison between LSI-based and VSM-based techniques at threshold of 0.3

Domain	No. of clusters	No. of real similar clusters	LSI-based		VSM-based	
			No. similar cluster correct/total	F-measure	No. similar cluster correct/total	F-measure
Spicy	2/1	2	2/2	1.0	1/1	0.67
University	1/2	2	2/2	1.0	1/1	0.67
Web	4/3	4	4/4	1.0	4/4	1.0
TPC_H	6/1	4	3/3	0.867	1/1	0.4
Finance	1/2	2	2/2	1.0	2/2	1.0
GeneX	10/8	12	11/14	0.85	9/10	0.8
Mondial	10/10	11	11/13	0.92	11/81	0.23
PO(large)	57/56	80/112	100/112			

of the LSI-based method to discover the hidden semantic relationships between schema element names.

Effect of LSI-Based on Matching Quality. The second set of experiments has been conducted to study the effect of both LSI-based and VSM-based methods on the matching quality. After selecting the "best" similar clusters, we first applied the name matcher on each matching task using data sets shown in Table 1. We then applied both the name and type similarity measures on the same matching task. Each task produces a subset of the match result. These subsets are then combined to generate the final match result. The final match result is evaluated using evaluation criteria, including precision, recall, and F-measure. Results for the matching quality are reported in Fig. 6.

The figures show, in general, that the LSI-based method has higher matching quality than the VSM-based method. This fact can be observed for the *University*, *Spicy*, *TPC_H*, and *PO* schemas. This can be clarified as these schemas have a high degree of semantic heterogeneity, and the LSI-based method has the ability to discover hidden semantic relationships between schema elements. However, the *Web*, *Finance* and *Genex* have less degree of heterogeneity, which results in nearly equal matching quality by both methods. In the case of the *Mondial* schema, which contains nearly no semantic heterogeneity, a large number of false positives are produced by the VSM-based method. Therefore, the LSI-based method produces higher matching quality than the VSM-based method w.r.t. the *Mondial* schema. It should be noted that the name matcher is more effective than the type similarity measure, and using the type measure makes a slight improvement in the matching quality.

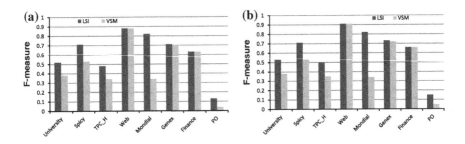

Fig. 6. Match quality comparison.

6   Conclusions

Partitioning-based techniques have become well-known approaches to match large schemas and ontologies. It has been proven that they improve the matching efficiency, however, they do not guarantee the matching quality. Identifying similar partitions of two schema trees is a crucial step before the matching process. To this end, in this paper, we introduced a new approach to cope with the problem of similar cluster determinations in the context of matching large-scale schemas. The proposed approach captures the features introduced by the Latent semantic indexing scheme to discover hidden semantic relationships between two sets of clusters. We in particular developed a matching framework focusing on the similar cluster determination step. Input schemas are first parsed and represented internally as schema trees to make the matching framework more generic. We then applied a clustering algorithm to partition each schema tree into a set of clusters. Further, we introduced a LSI-based algorithm to identify and determine similar clusters. To validate the performance of the proposed approach, we conducted a set of experiments utilizing different data sets comparing it with the classical vector space model (VSM)-based approach. The results proved that the LSI-based method outperforms the VSM-based method in determining the most similar clusters. It has the ability to discover hidden semantic relationships between schemas' elements. Therefore, the LSI-based method produces better matching quality than the VSM-based method. In future work we plan to extend the framework to explore the effect of the LSI-based method on matching efficiency. We need to validate some optimization techniques to enhance the LSI-based method.

Acknowledgments. This paper is a revised and extended version of the paper presented in [26]. A. Algergawy partially worked on this paper while at Magdeburg University.

References

1. Abiteboul, S., Suciu, D., Buneman, P.: Data on the Web: From Relations to Semi-structed Data and XML. Morgan Kaufmann, San Francisco (2000)
2. Algergawy, A., Massmann, S., Rahm, E.: A clustering-based approach for large-scale ontology matching. In: Eder, J., Bielikova, M., Tjoa, A.M. (eds.) ADBIS 2011. LNCS, vol. 6909, pp. 415–428. Springer, Heidelberg (2011)
3. Algergawy, A., Nayak, R., Saake, G.: Element similarity measures in XML schema matching. Inf. Sci. **180**(24), 4975–4998 (2010)
4. Algergawy, A., Nayak, R., Siegmund, N., Köppen, V., Saake, G.: Combining schema and level-based matching for web service discovery. In: Benatallah, B., Casati, F., Kappel, G., Rossi, G. (eds.) ICWE 2010. LNCS, vol. 6189, pp. 114–128. Springer, Heidelberg (2010)
5. Algergawy, A., Schallehn, E., Saake, G.: Improving XML schema matching using Prüfer sequences. DKE **68**(8), 728–747 (2009)
6. Aslan, G., McLeod, D.: Semantic heterogeneity resolution in federated databases by metadata implantation and stepwise evolution. VLDB J. **8**(2), 120–132 (1999)
7. Bellahsene, Z., Bonifati, A., Rahm. E.: Schema Matching and Mapping. Springer, Heidelberg (2011).
8. Berry, M.W., Drmac, Z., Jessup, E.R.: Matrices, vector spaces, and information retrieval. SIAM Rev. **41**(2), 335–362 (1999)
9. Bonifati, A., Mecca, G., Pappalardo, A., Raunich, S., Summa, G.: Schema mapping verification: the spicy way. In: EDBT 2008, France, pp. 85–96 (2008)

10. Chiticariu, L., Hernández, M.A., Kolaitis, P.G., Popa, L.: Semi-automatic schema integration in Clio. In: VLDB'07, pp. 1326–1329 (2007)
11. Cohen, W.W., Ravikumar, P., Fienberg, S.E.: A comparison of string distance metrics for name-matching tasks. In: IIWeb, pp. 73–78 (2003)
12. Deerwester, S., Dumais, S.T., Harshman, R.: Indexing by latent semantic analysis. J. Am. Soc. Inf. Sci. **41**, 391–407 (1990)
13. Do, H.H., Melnik, S., Rahm, E.: Comparison of schema matching evaluations. In: The 2nd International Workshop on Web Databases (2002)
14. Do, H.H., Rahm, E.: Matching large schemas: approaches and evaluation. Inf. Syst. **32**(6), 857–885 (2007)
15. Doan, A., Halevy, A.: Semantic integration research in the database community: a brief survey. AAAI AI Mag. **25**(1), 83–94 (2005)
16. Doan, A., Halevy, A.Y., Ives, Z.G.: Principles of Data Integration. Morgan Kaufmann, San Francisco (2012)
17. Ehrig, M., Staab, S.: QOM – quick ontology mapping. In: McIlraith, S.A., Plexousakis, D., van Harmelen, F. (eds.) ISWC 2004. LNCS, vol. 3298, pp. 683–697. Springer, Heidelberg (2004)
18. Halevy, A.Y., Ives, Z.G., Suciu, D., Tatarinov, I.: Schema mediation in peer data management systems. In: 19th International Conference on Data Engineering, pp. 505–516 (2003)
19. Hamdi, F., Safar, B., Reynaud, C., Zargayouna, H.: Alignment-based partitioning of large-scale ontologies. In: Guillet, F., Ritschard, G., Zighed, D.A., Briand, H. (eds.) Advances in Knowledge Discovery and Management. SCI, vol. 292, pp. 251–269. Springer, Heidelberg (2010)
20. Hao, Y., Zhang, Y.: Web services discovery based on schema matching. In: ACSC 2007, pp. 107–113 (2007)
21. Hu, W., Qu, Y., Cheng, G.: Matching large ontologies: a divide-and-conquer approach. DKE **67**, 140–160 (2008)
22. Landauer, T.: Handbook of Latent Semantic Analysis. Lawrence Erlbaum, Mahwah (2007)
23. Lee, D., Chu, W.W.: Comparative analysis of six XML schema languages. SIGMOD Rec. **9**(3), 76–87 (2000)
24. Lee, M.L., Yang, L.H., Hsu, W., Yang, X.: Xclust: clustering XML schemas for effective integration. In: CIKM'02, pp. 63–74 (2002)
25. Manning, C.D., Raghavan, P., Schutze, H.: Introduction to Information Retrieval. Cambridge University Press, New York (2008)
26. Moawed, S., Algergawy, A., Sarhan, A., Eldosouky, A., Saake, G.: A latent semantic indexing-based approach to determine similar clusters in large-scale schema matching. In: Catania, B., et al. (eds.) New Trends in Databases and Information Systems. AISC, vol. 241, pp. 267–276. Springer, Heidelberg (2014)
27. Peukert, E., Berthold, H., Rahm, E.: Rewrite techniques for performance optimization of schema matching processes. In: EDBT, pp. 453–464 (2010)
28. Peukert, E., Eberius, J., Rahm, E.: A self-configuring schema matching system. In: 28th International Conference on Data Engineering (ICDE), 2012, pp. 306–317 (2012)
29. Peukert, E., Massmann, S., Konig, K.: Comparing similarity combination methods for schema matching. In: GI-Workshop, pp. 692–701 (2010)
30. Rahm, E.: Towards large-scale schema and ontology matching. In: Bellahsene, Z., Bonifati, A., Rahm, E. (eds.) Schema Matching and Mapping. Data-Centric Systems and Applications, pp. 3–27. Springer, Heidelberg (2011)

31. Rahm, E., Bernstein, P.A.: A survey of approaches to automatic schema matching. VLDB J. **10**(4), 334–350 (2001)
32. Seddiquia, M.H., Aono, M.: An efficient and scalable algorithm for segmented alignment of ontologies of arbitrary size. Web Semant. **7**(4), 344–356 (2009)
33. Shvaiko, P., Euzenat, J.: Ontology matching: state of the art and future challenges. IEEE Trans. Knowl. Data Eng. **25**(1), 158–176 (2013)
34. Thuy, P.: Hybrid similarity measure for XML data integration and transformation. Ph.D. thesis, Seoul, Korea (2012)
35. Wang, Z., Wang, Y., Zhang, S.-S., Shen, G., Du, T.: Matching large scale ontology effectively. In: Mizoguchi, R., Shi, Z.-Z., Giunchiglia, F. (eds.) ASWC 2006. LNCS, vol. 4185, pp. 99–105. Springer, Heidelberg (2006)
36. Zhong, Q., Li, H., Li, J., Xie, G.T., Tang, J., Zhou, L., Pan, Y.: A Gauss function based approach for unbalanced ontology matching. In: ACM SIGMOD International Conference on Management of Data, (SIGMOD 2009), pp. 669–680 (2009)

Author Index

Printed in the United States
By Bookmasters